A **CLIMATE VOCABULARY** OF THE **FUTURE**

HERB SIMMENS

A Climate Vocabulary of the Future

Published by Wheatmark®
2030 East Speedway Boulevard, Suite 106
Tucson, Arizona 85719 USA
www.wheatmark.com

ISBN: 978-1-62787-508-0 (paperback)
ISBN: 978-1-62787-509-7 (ebook)
LCCN: 2017942522

In our everyday experience, if something has never happened before, we are generally safe in assuming it is not going to happen in the future, but the exceptions can kill you, and climate change is one of those exceptions.

—Al Gore

Carbon and methane are seizing physical territory, sowing havoc and panic, racking up casualties, and even destabilizing governments ... It's not that global warming is *like* a world war. It *is* a world war.

—Bill McKibben

We're the first generation to feel the impact of climate change and the last generation that can do something about it.

—Washington Governor Jay Inslee

To: Rosalie
Please enjoy this book on climate
Thank you

Contents

Introduction

It's 2025.

After a few relatively quiet years, later dubbed the *Clitanic Era,* where much of the evidence of climate change was literally or figuratively hidden below the surface, all hell broke loose.

Carbon forests are in flames, leading to *fire pollution* almost everywhere. *Bog burns* cover vast areas of the planet.

These conflagrations, which resulted from record breaking temperature-spikes and dry ground, produced *carbon clogs* in many forests. The Amazon, in particular, became a *source* of carbon instead of a carbon *sink,* the opposite of what usually happens. *Hyperfeedbacks* resulting from *clogs, albedo* decline and *permadeath* led to chaos and destruction.

Millions have become *climate refugees,* following the *first flee/ers.* A frantic *biotic migration* is underway, with marine and terrestrial life heading to cooler climates. Unfortunately, most will not make it, casualties of *biotic genocide.* An estimated five million people became victims of *carbicide,* some from *heat harm or drunken trees,* many from *sunny-day floods* or *zombie bacteria* that emerged out of hibernation from permafrost, as had happened in Siberia in 2016.

Most people were shocked that this could happen, experiencing an *oh shit moment* of dread, a profound *ecocholia.* After all, only a third of the country had been alarmed or even concerned about climate change, or what is increasingly known as *nuclear summer,* according to the well-known *Six Americas* survey.

Carbon recriminations dominated public discussions. TV and radio, where a *media omertà,* or code of silence about climate change, was angrily blamed. Also vilified was the *denial triad,* those who for reasons of religion, ideology or economic interest disputed the reality of climate change.

Everyone demanded action. The *ameliorists* felt that modest efforts were sufficient; the *eemergency* activists demanded a worldwide climate mobilization; the *light green environmentalists* were satisfied if free LED lights were given to all.

Intentional grounding of all remaining fossil fuels within three years was asked for by the *carbon supply-siders*. No one, however, could figure out how to prevent *carbon piracy* within the soon-to-be created *carbon exclusion zones* from sabotaging the proposed *Keep It in the Ground* policy.

In desperation, a small group of international leaders formed what they not-so-modestly called *The Climate Committee to Save the World* (CCSW). The first thing they did was consult with the world's leading experts on *climate chaos* — the *carbon bariatricians, planetarians, planitects, climate connectomics* practitioners *and climademics.*

Next the CCSW met with the *Dirty 90*, the companies that had extracted the most fossil fuels since the 1700s. The 90 reluctantly agreed to develop a *carbon retirement* program.

Opposition to *carbon retirement* soon emerged, led by a recently-formed interest group, *Friends of the Enemies of the Earth.* Once the *carbon police* heard about the actions of *Friends* they filed charges accusing them of being *carbon war criminals and* were handed over to the newly-created *climate justice system.*

This led to so much media attention that for the first time in a generation the *Kardashian Climate Index* showed more people searching online for 'climate change' than for Kim and her cohorts.

The *carbon bariatricians* then performed their *carbonoscopies* and reported back to the CCSW. The plan they came up with was brilliant and audacious and looked like it could save the world. Unfortunately, it soon succumbed to the dreaded *frozen chicken syndrome.* And that was the end of that.

That's what the words and phrases (shown in italics) of a *Climate Vocabulary of the Future* (*CVF*) look like. The above scenario is not a forecast. It is only meant to be illustrative of one possible *post normal climate.*

The *CVF* goes beyond dry definitions, encompassing both the technical and the conversational. It suggests new ways of thinking and speaking, and presents ideas for action needed to navigate this most uncertain of futures.

The vocabulary is informed by an understanding that the world is facing a true planetary *eemergency* — yes, the word does have two *ee*'s at the beginning. (See the entry on page 135 for more.) It is a mostly silent eemergency, as surveys suggest that two-thirds of Americans are either not aware of, or actively deny, the dire situation we face.

Following the trajectory of climate change as the director of New Jersey's statewide planning office for a decade, then as head of a climate change related nonprofit organization, and now as a writer and activist, I have gone from viewing climate change with curiosity to concern, to worry, and now to alarm.

Much evidence suggests that the planet's climate is near one or more tipping points that could easily lead to catastrophic climate change. While no one can know exactly what that will look like, when it occurs (many argue that it has already begun), or how devastating its effects may be, we do know it will last for eons, and put the habitability of the planet in grave danger.

My research, unfortunately, left me with no easy optimism. I kept looking for the mirage in the climate desert, the helpful negative feedback loop, the overlooked error that might ease my angst. I'm still looking.

My original intent was to curate a climate vocabulary, not a *climate vocabulary of the future*. As I examined the climate terminology, however, I saw large gaps, even chasms, in the current climate language. Rather than attempting to ignore or vault over these chasms, I began filling them in.

As Bill McKibben, author and preeminent climate activist, has said, "Yes we need new wind turbines and solar panels. But really, most of all we need new metaphors."

Lera Boroditsky, a linguistic philosopher at UCSD, states:

> Research shows us that the languages we speak not only reflect or express our thoughts, but also shape the very thoughts we wish to express. The structures that exist in our languages profoundly shape how we construct reality.

It is my hope that the words, phrases and metaphors in the *CVF* can help shape our thoughts and our actions.

The *CVF* is broad-based, emphasizing the behavioral, moral, and natural, as well as the economic and scientific dimensions, of climate change. Many of the terms cross disciplinary boundaries just as climate change crosses our planet's boundaries.

Over 450 entries are included, with 185 important and often misunderstood currently-used terms, along with 265 newly-created terms or neologisms.

The goal of the *CVF* is to encourage their use by escaping the confines of your laptop or the pages of your book.

How? Perhaps by hitching a ride on a blog, being mentioned in a conversation or noticed in a slide presentation a scientist or citizen is showing. Another way to think of it is to imagine these words as an invasive species coming to replace the old, tired and familiar climate vocabulary.

And much of our climate vocabulary *is* tired. A lot of what we know about climate change has changed during the past 25 years, except for the language we use to describe it.

The anodyne nature of some commonly-used terms (*climate change, global warming*), the inaccessibility of others (*mitigation, forcing mechanisms, anthropogenic*) and the word salads that characterize important concepts (*Intended Nationally Determined Contributions, Common but Differentiated Responsibility in Light of National Circumstances with Respective Capability*) often obscure their meaning and importance.

Even terms that appear to be clear are often not. Wikipedia's description of *runaway climate change*, for example, indicates that, "runaway climate change is hypothesized to follow a tipping point in the climate system, after accumulated climate change initiates a reinforcing positive feedback". Got that?

The *CVF* attempts to freshen the climate vocabulary by painting a vivid picture of a trend, behavior, need, state of mind, technology or scientific phenomena. The entry may be a straightforward word or phrase like *dark snow* or *negative emissions*, a playful term like *frozen chicken syndrome* or *robin carbon hood tax* or a sharply-drawn phrase like *carbon war criminal* or the *no solutions coalition*.

While many of the entries reflect a distinct perspective, the *CVF* is careful to be factual. As Senator Daniel Patrick Moynihan said, "Everyone is entitled to his own opinion, but not his own facts". And there's no special entitlement here.

You might shake your head at an occasional snippet of sarcasm or cynicism embedded in an entry; it's just my defense mechanism against climate despair, a linguistic device designed to allow me to dive even deeper.

If you're unfamiliar with the implications of the *carbon budget* or the likely consequences of *climate positive feedback*, reading this book could be more than a little unsettling. You may even experience, as I have, a transformative *oh shit moment*.

If you've now decided to put the book down or power off your Kindle as you have little appetite for gloom, know that the *CVF* contains about 70 novel ideas and proposals. These proposals briefly describe new organizations, technologies or processes to make life better and to help keep climate change from getting worse.

Using A *Climate Vocabulary of the Future*

The *CVF* is organized in three sections. First, are words and phrases in current use. These are followed by newly-formulated terms or terms that are used for the first time (according to Google anyway) in the context of climate change. Original proposals for improving life or fighting climate change make up the last section.

A **bold** word or phrase used in the body of an entry signifies a word or phrase that has its own entry.

In many entries *climate chaos* is substituted for 'climate change' as chaos vividly — and more accurately — expresses the emerging reality of climate change. In some cases, 'global heating' or 'global scorching' is substituted for *global warming* for the same reason.

Temperatures are given in Fahrenheit, the temperature scale used in the United States. To those of us who think in Fahrenheit, a 2° Celsius increase sounds a lot less threatening than a 3.6° Fahrenheit increase. And the last thing we need is to continue to understate the climate change threat.

Herbsimmens.com has been created to highlight new and updated entries. Please visit. I also welcome your climate vocabulary ideas. Kindly send them to hsimmens@gmail.com.

The Climate Crisis and What to Do About It

The entries in the *CVF* will have the most impact if linked to a plan of action to tackle climate chaos. And while describing an action blueprint in detail goes well beyond the scope of this book, here's my perspective on how we got here and what we need to do.

The climate crisis was not inevitable. Vigorous action could have been taken in the 1990s and the 2000s to transform our economy, the most carbon intensive in the world. (In fairness, President Clinton did propose an energy tax in 1993.) We were also in a position to lead the world by example. Had we done so, the planet would have escaped some, perhaps much, of the climate chaos that has occurred since then.

But we didn't act — for both benign and malign reasons. Perhaps we could have been excused for having a failure of imagination that it would be this bad, or because we thought we had more time to act.

And we can hardly be blamed (though perhaps the Supreme Court can be) for the cosmically tragic luck that denied climate hawk Al Gore the Presidency, largely because of a poorly-designed ballot in Florida. (And how can we even begin to make sense of an FBI director and Russian hacking interfering in the 2016 election, thus helping to elect a climate denier.)

The cowardice, ignorance and avarice of many of our political leaders and corporate titans have been largely responsible for our collective inaction. The oil companies in particular appear to have known about the dire risks of carbon combustion for half a century but they — like the tobacco companies with human health — have put their narrow interests ahead of the planet's health.

The indifference (*media omertà*) of our mass media and our consumerist addiction (itself partly a function of the billions of advertisements that our media serves up) further explains a lot of our climate passivity.

Nor do we have a climate Churchill, Mandela or Gandhi, an FDR or Martin Luther King, Jr. to inspire, lead and unite our society.

The absence of both leadership and a media that urgently explains the emergency we face means that we as citizens need to take the lead in demanding immediate action. And that action is the establishment of a climate emergency mobilization on the scale of the Second World War, with the simple and understandable goal of decarbonizing the country's economy and the planet's economy within a generation, as climate science requires. I call it the One Generation Challenge.

Whether or not such a mobilization ultimately succeeds, the solidarity resulting from bringing our country and the world together can help us survive the harsh times and heartbreaking climate catastrophes awaiting us.

Climate mobilization will be particularly critical for the young, who have the most to lose, if we do nothing. These generations will be most inspired by the opportunity to be part of a truly heroic effort.

The *central climate conundrum* is why we haven't yet mobilized, knowing that we're in the process of destroying our own lives and the lives of virtually every species on earth.

Maybe the answer to the conundrum is simpler than we think. To paraphrase what former Texas Governor Ann Richards said about George W. Bush, "We in the modern world were born on third base and yet thought we hit a triple".

It was our fortuitous (at least until now) discovery of fossil fuels more than our cleverness, hard work, technology, religion or economic system that placed us on third base. And now the planet is in the process of throwing us out while we attempt to steal home.

I hope *A Climate Vocabulary of the Future* helps you to reach home as you navigate the *post normal climate* that we are all beginning to experience.

Current Climate Terms

The following are some of the more common, important and interesting terms in use. This is not a complete list of climate change terms — there are thousands!

Adaptation/Mitigation

These terms are perhaps the most familiar in the climate chaos lexicon. It is said that adaptation refers to actions designed to protect us from the earth, and mitigation refers to actions designed to protect the earth from us.

Adaptation anticipates the adverse effects of climate chaos and takes appropriate action to prevent or minimize these effects. Well-planned early adaptation saves money and lives.

Adaptation (or **readaptation**) may consist of such physical actions as elevating buildings to withstand floods, detecting water leaks to provide more usable water, or the anticipatory relocation of part or all of a city or even a nation to higher or safer land. Safety drills, psychological counseling and community organizing to build social capital between neighbors are a few of the many 'soft' adaptation interventions.

An often overlooked and critical dimension of adaptation is the need to assist plants and animals to thrive as the climate warms. Many plants and animals are migrating to colder climates at remarkable speeds; some are shrinking to better manage the heat, while others are changing their metabolism. Scientists are applying genetic engineering and other techniques to enable at least some plants and animals to survive a **post normal climate**.

Mitigation is the prevention or minimization of climate chaos. Unfortunately, the world is past the point where full mitigation is possible.

See Biotic Refugees, Carbon Tourniquet, Climate Palliation, Climate Shift-shaping, Sources and Sinks, Transplant Nations

Agroecology

The artful application of ecological principles to the science and practice of agriculture.

Agricultural practices are both a major cause and effect of climate chaos. A significant though hotly-disputed percentage of carbon emissions results from agricultural processes. Methane emitted from cattle, sheep and rice paddies and CO_2 emitted from the pervasive mechanization of farming contribute to the carbon footprint.

While agricultural experts and farmers recognize the urgent need to adapt farming to hotter, wetter, dryer, stormier, 'pestier' and just plain weirder weather, while also reducing carbon emissions, the direction this adaptation should take is vigorously debated.

The options open to restructure agriculture — while varying greatly depending on region, country, crop and potential climate impacts — are many.

The business as usual (BAU) approach relies on land, crop and animal intensification supported by the greater application of chemical inputs, mechanization and genetic modification.

Unlike BAU, agroecology focuses on using natural systems to augment agricultural productivity and reduce and drawdown emissions. Among the more promising agroecology approaches is the cultivation of perennial rather than annual crops, as advocated by the Land Institute in Salina, Kansas.

'The small is beautiful' philosophy argues for the superiority of organic, biodynamic and small-scale family farming, often mixing animal grazing and vegetable farming in creative ways known as permaculture.

Many foundations, national governments and agrochemical companies will be pitted against local farmers and idealistic agrinauts as the battle for the future of **post normal climate** agriculture plays out.

See 4/1000 Initiative, Plant Purgatory, Savory Holistic Range Management, Terra Preta

---

Albedo

The proportion of light that is reflected from a surface.

The lighter the color, the higher the albedo; the darker the color, the lower the albedo.

Less ice and snow and more melt water means a lower albedo, with less sunlight reflected away from the earth and more remaining to heat the atmosphere and the surrounding water.

Recent studies show that the warming of the earth has increased by about 25% more than it otherwise would have as a result of the extra heat generated by a lower albedo.

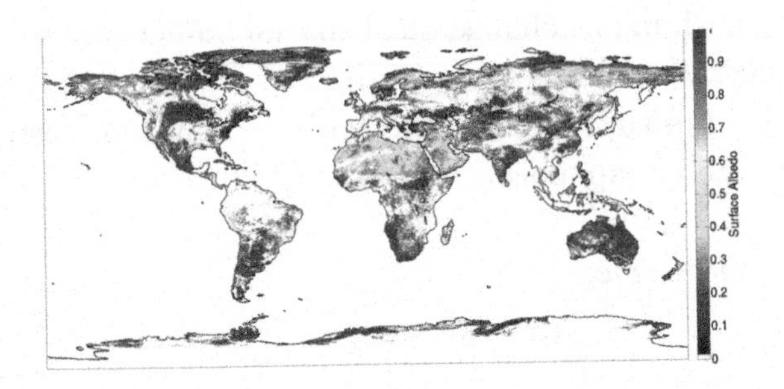

This process is an example of a **climate positive feedback** where sea ice loss leads to albedo decline, which leads to greater heat absorption, and then to further sea ice loss until all ice has melted, as is happening in the Arctic.

See Arctic Amplification, Climate Sensitivity, Dark Snow, No Ice Age, Watermelon Snow, White Only Laws

http://neven1.typepad.com/blog/2014/02/decreasing-arctic-albedo-boosts-global-warming.html

---

Antemortem

An activity performed prior to death. In the climate context, an antemortem would examine why it appears that we are not able to prevent the death and destruction of modern civilization. Antemortems may be particularly valuable as few people may be alive to prepare postmortems.

See Biomortis, Central Climate Conundrum

Anthropocene

A term proposed by the ecologist Paul Crutzen in 2000 describing the current geological era. The defining characteristic of the Anthropocene is the decisive influence by humans on the earth's natural and environmental characteristics, including climate.

The arrival of the Anthropocene anoints *homo sapiens* as responsible for the health and well-being of our planet. Will naming a geological era for humans change our behavior sufficiently so we finally become effective stewards of the earth?

And will humanity survive the Anthropocene era, particularly in the time of the Trumpopocene?

See Psychopocene

Arctic Amplification

The observed phenomenon that climate chaos related atmospheric warming is about two to three times greater in the Arctic than in temperate and equatorial zones.

The loss of sea ice appears to be the primary factor in explaining this phenomenon. Because of global weather interconnections, arctic amplification is believed to affect weather patterns in other parts of the globe, including such disparate events as the recent California drought and Superstorm Sandy in 2012.

The consequences of Arctic amplification are potentially catastrophic given the likelihood that higher temperatures in the Arctic will lead to the release of vast quantities of CO_2 and methane as a result of thawing permafrost, and to further warming due to sea ice melting caused by **albedo** decline.

See Climate Positive Feedback, Permadeath, Tipping Point, Warm Arctic Cold Continents, Zombie Bacteria

https://unfccc.int/files/science/workstreams/research/application/pdf/5_wgiar5_hezel_sbsta40_short.pdf

Autotrophic Cities

Cities that are largely self-contained and self-reliant.

Communities face being cut off from their hinterlands due to natural or man-made disasters. Climate chaos related limitations on the production and transportation of goods and services can also isolate cities. As a result, cities will need to become more self-reliant. As they convert to **circular economies**, their food, wastes and material systems become increasingly localized, while their economies become more resilient.

See Photosyncities

Backfire Effect

A backfire effect occurs when energy savings are cancelled by greater consumption due to the extra income from energy savings.

Another kind of backfire effect occurs when many climate deniers or skeptics receive persuasive information about the reality of climate change. Not only are they rarely persuaded by the information they receive, they become more entrenched in their denialist beliefs. This is an important, albeit highly unfortunate, phenomenon that deserves attention by those who believe that with the right message deniers will stop denying.

See Rebound Effect

http://www.realclimate.org/index.php/archives/2016/06/boomerangs-versus-javelins-the-impact-of-polarization-on-climate-change-communication/#more-19502

Bending the Climate Curve

Reducing the magnitude and pace of global temperature increases that will result if we continue with business as usual.

Bending, if not **breaking the climate curve**, is essential to maintain a habitable planet.

Bioenergy with Carbon Capture and Storage (BECCS)

A process that removes CO_2 from the air through cultivating plants, burning them and capturing and sequestering the emitted CO_2.

Plants are cultivated and then burned for energy. The waste CO_2 is then captured, liquefied, transported and stored underground in secure locations for indefinite periods.

Perhaps more than any other technology, BECCS is considered by many as the Holy Grail in its potential to **bend the climate curve**. Yet BECCS is a relatively new process, with few facilities in operation.

BECCS is intended to result in negative emissions, meaning that if designed and operated as expected, the process will take more carbon out of the atmosphere than it adds while generating usable energy to power civilization.

This climate dividend results from the extra carbon absorbed from the atmosphere by plants as they are grown for use as bioenergy.

Significant technical and economic problems associated with BECCS include the high-energy cost and low **EROEI** of fueling BECCS, the complexity of building the required systems at the scale required, the challenge of growing and cultivating the millions of acres of biomass without raising food prices, and the need to locate thousands of sites to safely sequester CO_2 for many generations — all this in a **post normal climate** characterized by political and economic disruptions, mass migration and the likelihood of agricultural decline and widespread climate-induced disease.

Like **geoengineering**, BECCS appears to be an elegant solution that allows us to continue to burn carbon. But endorsing it as a distant game changer is like a baseball team counting on a pitcher who has just been drafted to win the 7th game of the World Series a decade from now.

See CCS, EROEI, Negative Emissions

https://hub.globalccsinstitute.com/publications/global-status-beccs-projects-2010/2-scientific-background-beccs

---

Biofuels

Liquid fuels derived from living matter such as corn or algae.

Proponents argue that use of these fuels reduces carbon pollution and keeps fossil fuels in the ground while providing markets and incomes to farmers and small businesses.

Critics emphasize the significant negative effects, including an increase in food costs, food shortages and even starvation, as well as increases in carbon emissions over the full biofuel life-cycle.

The use of biofuels could grow rapidly if BECCS facilities become widespread.

See BECCS, CCS

Bio Geoengineering

Applying technological processes to biological systems with the goal of capturing CO_2 from the atmosphere or preventing the release of CO_2 into the atmosphere.

Examples include increasing tree cover, changing grazing patterns and biochar soil enrichment.

See Carbon Forests, 4/1000 Initiative, Savory Holistic Range, Carbon Drawdown, Management, Terra Preta

Biophilia

Coined by the naturalist Edmund Wilson to describe an innate caring and feeling that humans have for the natural world.

While an appealing and intuitively resonant idea, if biophilia is truly innate and universal, why then are we ravaging the natural world in so many ways?

This is the **Central Climate Conundrum**.

Bio Sequestration

The removal of CO2 from the atmosphere by natural processes, (which may be assisted by humans) primarily photosynthesis, in contrast to **Chemo Sequestration**.

See Bio Geoengineering, Carbon Forests, 4/ 1000 Initiative, Terra Preta

Biotic Genocide

The large-scale decimation of terrestrial and ocean life resulting from human influences, particularly pollution, habitat destruction and climate change.

Blue Carbon

Carbon stored in sea grass, salt marsh, mangroves and other coastal ecosystems.

Large quantities of carbon are sequestered in these coastal ecosystems. These biologically rich areas are being lost to development at a very high rate. Their preservation is essential for food, flood protection, biodiversity and climate stability.

See Marine Heat Waves, Ocean Hawks

Boston Carbon Party

A term introduced by the writer Jeff Goodell in an interview with the famous climate scientist James Hansen in Rolling Stone magazine, December 2016.

He was referring to Hansen's comments that we need a new political party and a new energy system.

Bright, Dark and Light Green Environmentalism

A classification of environmental ideology and behavior based on the writings of the futurist Alex Steffen.

Bright Green Environmentalism argues that technologies and policies such as radical energy efficiency, city densification, electrification of energy production and consumption and electric self-driving cars can maintain and even enhance economic prosperity while convincingly addressing climate chaos and environmental degradation. This perspective relies on the wholesale transformation of our energy, transport, habitat, manufacturing and food systems.

Dark Green Environmentalism rejects technological and capitalist solutions to environmental problems and climate chaos. Only radical behavioral changes and political action that shrinks the consumer economy, supports back-to-nature movements and population limits, and focuses on voluntary simplicity in personal and communal life can adequately address climate chaos. It's similar in an American context to the European **degrowth** movement.

Light Green Environmentalism rejects the Dark Green Environmentalism opposition to capitalism and growth and is mostly indifferent to the high-tech transformational aspirations of Bright Green Environmentalism. The Light Greeners see an environmentalism that is a personal lifestyle choice as sufficient to insure environmental progress. Light Greeners tend to be **Brightsiders** and/or **Ameliorists.**

Brightsiders

Individuals who believe that climate chaos can be controlled without the necessity for **eemergency** action.

Brightsiders proclaim the brighter future that will occur as a result of a transition to renewable energies, reformed agricultural production and better urban design. Brightsiders are considered by many eemergency activists as essentially soft climate deniers or at least as **climate gradualists**. As climate chaos deteriorates, it is hard to imagine that many will remain brightsiders.

See Ameliorists, Bright, Light and Dark Environmentalism, Denial Triad, Eemergency

---

Cap and Trade

A system of regulating carbon emissions used in Europe, several states in the Northeast and California among other places. About 20-25% of all atmospheric carbon emissions globally are governed by a cap and trade program.

A governmental body sets a limit or 'cap' on carbon emissions and requires emitters to obtain a permit either for free or by auction. If the emitter needs to emit more than the permit allows, he or she can purchase or trade permits from or with other permit holders.

Cap and trade is usually compared to a **carbon tax**, **fee and dividend** or cap and dividend as a mechanism to reduce carbon emissions. There are distinct advantages and disadvantages to each approach. The **climate hawk** community is divided over which system is more politically feasible, technically manageable and most likely to significantly reduce carbon emissions.

Carbon Budget

The amount of CO_2 that can be burned without going over a given worldwide temperature threshold.

The number of years until the planet exceeds the carbon budget depends on the chosen temperature threshold and the probability of remaining under that threshold. For example, the earth would have 77 years left for a 33% chance of remaining below 4.8° whereas there would only be six years left for a 66% chance of remaining below 2.7°.

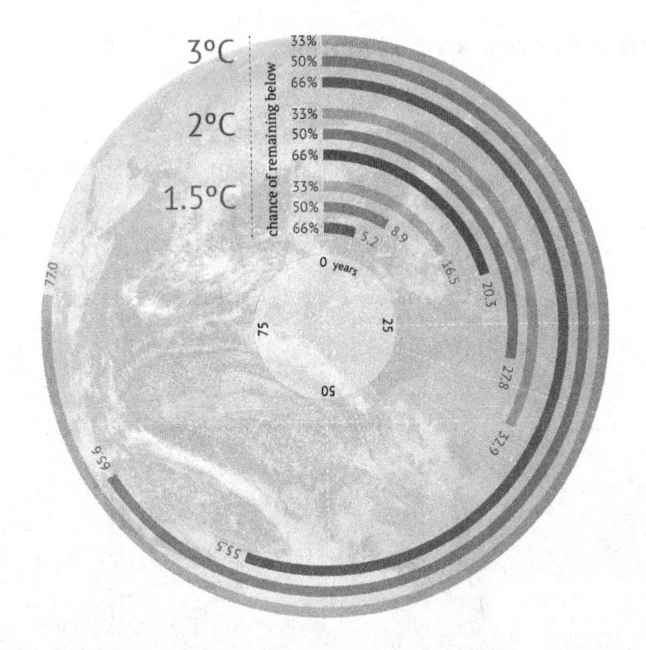

What's not clear from the literature is how many years we would have left with close to a 100% chance of remaining below 2.7°. This is the only option that provides a high level of confidence that we can maintain a reasonably habitable planet. The answer appears to be that it's already too late.

To have a very high probability of remaining under 2.7° or even 3.6° requires cutting carbon emissions on the order of 10% or more per year worldwide. This is close to impossible without creating a severe depression or a global climate panic that utterly transforms international climate politics. And many argue that 3.6° or perhaps even 2.7° is still unacceptably high as climate chaos would be extreme.

The carbon budget math demonstrates to most the necessity of a **One Generation Challenge** to decarbonize within a few years through a worldwide WW2-like climate mobilization.

See One Generation Challenge. 2° Dogma

Carbon Capture and Storage

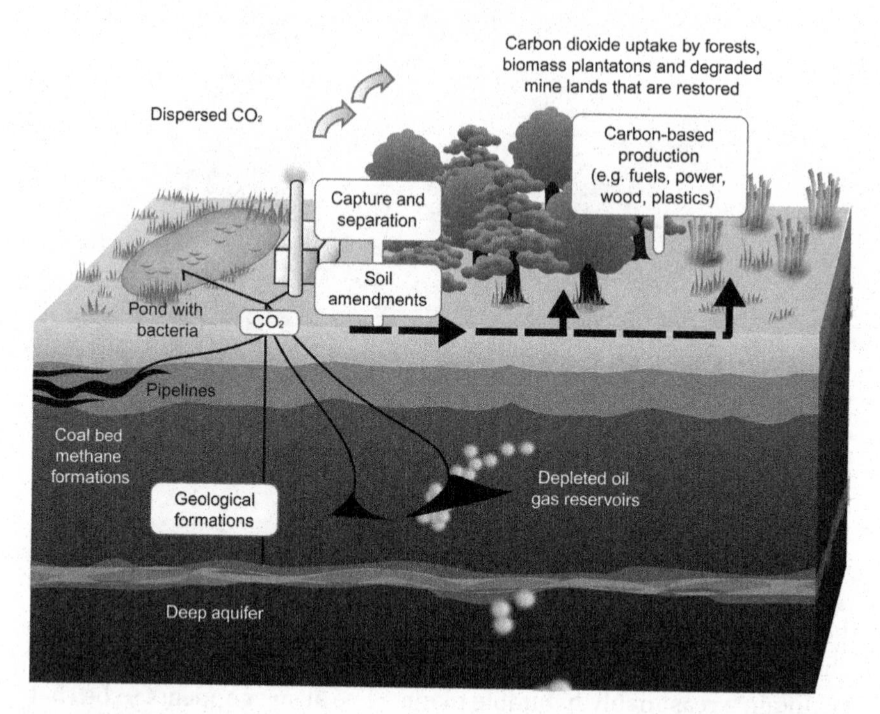

CCS is an advanced set of technologies that captures CO_2 from fuel combustion, transports the CO_2 largely by pipelines, and stores the CO_2 underground in former oil and gas fields and other geologically secure formations.

Currently, there are only a dozen or so relatively small CCS plants in operation throughout the world.

A successful CCS program would allow coal to continue to be burned without increasing the amount of CO_2 in the atmosphere (**carbon neutrality**). Critics are concerned that CCS will act as a kind of **moral license** to encourage coal to be burned for decades.

Scaling-up and cost are considered to be the two biggest problems assuming the technology gets perfected. As much as 25% to 45% more energy is required to operate a CCS facility than a conventional coal-burning power plant.

Thousands of miles of pipelines, scores of CCS plants and thousands of carefully sited deep wells would have to planned, designed, permitted and carefully operated throughout the planet in a very short time frame. The required infrastructure would exceed the current oil industry infrastructure.

Where would we get the materials, the sites, the technical experts, the capital and the political sophistication to beat back the **numby** protests? Who can believe that this is do-able in a generation or even two?

See BECCS, One Generation Challenge

http://www.iea.org/topics/ccs/
http://www.resilience.org/stories/2016-11-22/trump-s-coal-delusions

---

Carbon Combustion Complex

The constellation of industries that produce and/or benefit from the burning of carbon. Coined by Professor Naomi Orestes, it refers to oil and coal companies as well as to gas stations, distributors and the many related industries that depend on fossil fuels for economic vitality. This carbon combustion complex is fighting fossil fuel limitations with its vast economic and political influence.

See Denial Triad

Climate Craps

Well, not exactly craps, but it's a better title than 'climate bets'. Several climate realists have challenged climate deniers to bet whether temperatures will increase in future years. It appears that few if any deniers have been willing, at least in public, to put their money behind their so-called beliefs.

Why **climactivists** don't routinely and publicly challenge prominent deniers to put up or shut up is puzzling. A real lost opportunity.

http://www.jamstec.go.jp/frsgc/research/d5/jdannan/betting.html

Carbon Dioxide Equivalent CO_2e

CO_2e (e for equivalent) converts the impacts of the non-carbon gases to CO_2 based on their relative global warming potential.

This allows for consistent and comprehensive measurement. While CO_2 is by far the most prevalent of the greenhouse gases, other gases such as methane, nitrous oxide and ozone as well as water vapor, chlorofluorocarbons, and several others, function as **greenhouse gases**.

Some controversy exists about the conversion ratios given the differing times each gas remains in the atmosphere and the relative ability of each gas to raise temperatures.

https://climatechangeconnection.org/emissions/co2-equivalents/

Carbon Drawdown

The removal of CO_2 from the atmosphere by natural or human means.

Drawdown is becoming a default solution to the climate crisis. Projections demonstrate there is a very high probability that a habitable earth will no longer be possible without significant carbon drawdown, given the likelihood of exceeding the world's **carbon budget**.

The science, technology, politics, economics and timing of drawdown will emerge as central issues throughout the world in the years ahead.

Project Drawdown is a research effort initiated by the **planetarian** Paul Hawken to analyze 100 or more methods of drawdown. *Drawdown* was published in 2017. Note that drawdown as defined by the Project is "that point in time when the concentration of greenhouse gases in the atmosphere begin to decline on a year-to-year basis".

See BECCS, Bio Sequestration, Chemo Sequestration, Direct Air Capture, 4/1000 Initiative

https://www.amazon.com/Drawdown-Comprehensive-Proposed-Reverse-Warming/dp/0143130447

Carbon Emissions

Global carbon emissions are around 10 billion tons (10 gigatons) annually. CO_2 emissions are about 36 gigatons annually as CO_2 weighs about three and a half times as much as carbon. Fossil fuel and cement emissions account for about 90% of human generated carbon emissions with land use changes accounting for about 10%. About a quarter of emissions accumulate in the ocean, 30% on land and the rest in the atmosphere.

Carbon concentrations in the atmosphere increased from about 278 parts per million (ppm) two hundred years ago, to over 410 ppm in 2017, with the rate of increase noticeably greater in recent years. If and until carbon declines to 350 ppm or less, climate chaos will continue to be with us.

See Keeling Curve

Carbon Footprint

The amount of carbon generated by an individual, group or object over a particular period. The average US CO_2 footprint of over 16 tons per capita per year is about twice that of China and the EU and almost 10 times greater than that of India.

See Arctic Ice Sensitivity, Consumption Based Footprinting, Contraction and Convergence

carbonfootprintofnations.com/content

Carbon Forests

Planting of trees for the primary purpose of **carbon drawdown** from the atmosphere.

The scale of additional forest needed to significantly offset carbon emissions is vast.

Climate positive feedbacks can reduce or even eliminate the carbon drawdown of forests. Planting forests on plains or grasses is likely to reduce **albedo** due to the darker color of forests versus the fields they replace, offsetting some of the carbon benefits. The carbon cost of supplies and labor for planting and maintenance will further reduce carbon benefits as will more frequent and intense wildfires. Some forests may not even grow back after wildfires. Any enhanced CO_2 absorption by carbon forests will also be partially offset by reduced CO_2 absorption by the oceans.

There appears to be no carbon forest free lunch.

https://insideclimatenews.org/news/21122016/california-forests-wildfires-climate-change

See Carbon Maze

Carbon Humanitarianism

Carbon humanitarianism is a term coined by the author Michael T. Klare to signify the humanitarian yet problematic benefits of fossil fuels for people in undeveloped and developing countries. The term is based on arguments made by former Exxon CEO Wes Tillerson to justify continued use of fossil fuels. Tillerson is also associated with the concept of 'energy poverty', a term that suggests the utility of more energy use.

While there is little dispute that access to fossil fuels can provide billions with the opportunity to live more comfortable, productive (at least in the modern Western sense) lives, these benefits come with a Faustian choice of short-term prosperity at the price of long-term carbon chaos.

http://www.tomdispatch.com/blog/175940/

Carbon Incumbency See Carbon Combustion Complex

Carbon Labeling

Displaying the amount of CO_2 generated by the production and distribution of the product or service being purchased or consumed.

While consumer carbon labeling was enthusiastically embraced earlier in the century, in recent years it has almost disappeared. Reasons include the cost and time of research and labeling, the absence of standardized protocols, and doubts about how useful the information is to consumers.

Carbon Lock-in

Two types of carbon lock-in (carbon path dependence) can be identified. The first might be called natural carbon lock-in. *Climate change appears to be largely irreversible for 1000 years even after emissions stop.*

There's also man-made carbon lock-in. This occurs when we build coal-burning power plants that have multi-decade useful lives or manufacture tens of millions of cars and trucks that last for years. And each time we build a new home, office, school, hospital or neighborhood and don't design it to emit zero carbon, we are locking in carbon emissions for decades.

Unlike natural carbon lock-in, there's nothing immutable about man-made lock-in. We can collectively choose to close the coal-fired power plants and take the oil-fueled cars off the road well before their useful life is up, recognizing that the financial loss resulting from stranded assets is a lot less than the cost of losing a habitable planet. We can mandate strict zero carbon standards for construction and require communities to densify and promote low carbon transport.

The largest and most controversial sources of carbon lock-in are the proven fossil fuel reserves held by oil companies, coal companies and national governments. These reserves are about five times the amount as can be burned while still having a chance of staying at or less than a 3.6° increase in temperature.

The **Keep It in the Ground** campaign is an attempt to ensure that these reserves are **intentionally grounded**. The fight over who will bear the cost of retiring these reserves could well be the most important political and financial battle in history.

http://www.carbontracker.org/report/carbon-bubble/
https://ourchangingclimate.wordpress.com/2016/08/09/climate-inertia/
http://www.pnas.org/content/106/6/1704.

Carbon Offsets

A mechanism by which an entity reduces its emissions through funding projects that are intended to increase CO_2 absorption.

These offsets are usually available from an activity or project located somewhere other than where the credits will be applied.

Offsets are justifiably controversial given the difficulties in monitoring the projects as well as ensuring that the offsets lead to carbon emission reductions that would not otherwise occur. Many view offsets as morally questionable as they essentially function as licenses for carbon generators to continue to pollute.

The current volume of carbon offsets is only a fraction of total carbon emissions.

http://www.nature.com/news/the-inconvenient-truth-of-carbon-offsets-1.10373

See Cap and Trade

Carbon Pollution See Carbon Emissions

Carbon Sequestration

The long-term storage of carbon to ensure that it does not enter the atmosphere.

While sequestration can take many forms, it generally means the injection and storage of carbon underground in secure geological formations. It can also refer to the storage of carbon by living creatures such as plants, trees and algae as well as storage in the soil or oceans.

See BECCS, Bio Sequestration, Chemo Sequestration, CCS, 4/1000 initiative, Savory Holistic Range Management, Terra Preta

Carbon Tax

A fee levied on fossil fuels to discourage carbon emissions and to transform an economy to renewable or at least lower carbon energy. Carbon taxes are not particularly popular, having been levied in only a handful of countries. In North America, only British Columbia has a functioning carbon tax. **Cap and Trade** programs are more commonly used to reduce carbon emissions.

See Fee and Dividend

Chemo Sequestration

The removal of CO_2 from the atmosphere by chemical processes such as scrubbers and mineral carbonation. As distinguished from **bio sequestration.**

Children's Climate Crusade

Lawsuits have been filed on behalf of children arguing that the government has an obligation to preserve air and water resources. The most prominent was filed in Oregon by Our Children's Climate Trust arguing that children are entitled to a stable climate.

Climate activists Bill McKibben and Naomi Klein called this case "The most important lawsuit in the world."

The federal government in defending itself has argued that the government has no obligation to preserve the air and water resources of this country. Yes, *read that sentence again*! Oh, and the fossil fuel industry joined with the Obama Justice Department.

A trial in federal court is scheduled for early in 2018. The plaintiffs will now be required to demonstrate that the defendant companies and the federal government knew about the harmful effects of climate change and chose to do nothing.

See Climate Trustee, 10,000 Climate Law Suits

http://www.ourchildrenstrust.org/us/federal-lawsuit/

Circular Economy

An economy where industrial production outputs are transformed into inputs with minimal or no waste or pollution, similar to how natural systems work.

Transitioning to a circular economy is likely to lead to significantly lower carbon emissions as well as less pollution and greater economic efficiency.

A recent study by the Club of Rome analyzing Sweden's economy concluded that applying circular economy principles could reduce emissions by 70% and increase employment by 100,000 by 2030.

https://www.theguardian.com/sustainable-business/2015/apr/15/circular-economy-jobs-climate-carbon-emissions-eu-taxation

---

Clathrate Gun Hypothesis

The theory that increases in sea temperatures may lead to the sudden increase of methane in the form of methane clathrates, or methane bound up in sea ice. The release could lead to the release of methane on such a vast scale that it triggers significant increases in global scorching — hence, the use of the word 'gun'.

What the actual threat of such a release is, and the magnitude and timing of any resulting temperature increases, is a matter of continuing research.

Clear Air Turbulence

There is some evidence that climate chaos is increasing the clear air turbulence airplane flights experience.

https://www.theguardian.com/world/2016/sep/11/cost-bumpy-flights-air-turbulence-global-warming-united-airlines

See Global Weirding

Clexit

An international movement organized to prevent the Paris Climate Agreement from being ratified. Building on the Brexit vote, the Clexit effort is led by a group of prominent climate chaos deniers, several of whom were involved in tobacco and fossil-fuel related industries.

The Clexit leaders hope to capitalize on the same anti-establishment, anti-expert sentiment as Brexit exploited. This is not meant to suggest that Brexit supporters are necessarily climate deniers.

(Note that the Paris Agreement was ratified on October 4, 2016, when the requisite number of countries approved the treaty. With the treaty ratified, it is not clear whether the Clexit movement will continue in some form.)

https://www.theguardian.com/environment/climate-consensus-97-per-cent/2016/aug/08/rejection-of-experts-spreads-from-brexit-to-climate-change-with-clexit

Cli-fi

A fast-growing genre of fiction whose subject is climate chaos. There are now over 100 college courses focusing on Cli-fi.

Cli-fi was first named by Dan Bloom:

> The 'cli-fi' name came to me as I was thinking of ways to raise awareness of novels and movies about climate change issues. I toyed with using such terms as 'climafic' or 'climfic' or 'clific.'

for the longer term of 'climate fiction.' But I wanted an even shorter term that could fit easily into newspaper and magazine headlines. So, using the rhyming sounds of 'sci-fi,' I decided to go with the short, simple to say, and simple to write 'cli-fi'. And the short term caught on worldwide, beginning on April 20, 2013 when NPR radio did a five-minute radio segment about 'cli-fi.' That was the beginning of its global outreach and popularity among academics, literary critics, journalists and headline writers.

cli-fi.net

Climactivist

A climate action activist.

Climatarium

"A public hall comprising devices for presenting global and localized interrelationships between climate and society; and for simulation of future effects on same; for purpose of furthering scientific understanding, public instruction, and measures for climate resilience."

A description from a workshop conducted by the Urban Design Lab at Columbia University.

http://urbandesignlab.columbia.edu/projects/educational-research-2/urban-ecology-studio-2/pier-26-new-york-the-climatarium-of-to-morrow/

Climate Capital Budget

Governments and other organizations with large capital expenditures will increasingly create climate capital budgets to identify and prioritize those expenditures directed towards climate **mitigation** or **readaptation**.

Climate Careers

Climate chaos will require a host of new skills and new job categories. Many traditional jobs will need to be restructured to address the demands of climate chaos.

The writer and sustainability consultant John Thackara has an apt description of some of the new climate related jobs that are being created:

> Its ranks include energy angels, wind wizards, and watershed managers. There are bio-regional planners, ecological historians, and citizen foresters.
>
> Alongside dam removers, river restorers, and rain harvesters, there are urban farmers, seed bankers, and master conservers. There are building dismantlers, office-block refurbishers, and barn raisers. There are natural painters, and green plumbers. There are trailer-park renewers, and land-share brokers.
>
> Their number includes FabLabs, hacker spaces, and the maker movement. The movement involves computer recyclers, hardware re-mixers, and textile upcyclers. It extends to local currency designers. There are community doctors. And elder carers. And ecological teachers.

See Carbon Bariatrician, Climademic, International Climate Corps, Planitarian, Planitect

http://www.resilience.org/stories/2016-12-12/from-oil-age-to-soil-age

Climate Change

The term commonly used to signify the consequence of **global warming**.

Resulting from adding greenhouse gases to the atmosphere, global warming leads to a variety of destabilizing and not always predictable climatic changes. The more global warming there is, the more the climate changes.

Climate Chaos

Arguably, a more accurate and vivid term for **climate change.**

Chaos because of the extreme unpredictability of the **post normal climate** and the extreme disorder resulting from these climate upheavals. The *CVF* generally employs climate chaos as a substitute for climate change.

Climate Cliff

The earth on the verge of reaching one or more **tipping points** that could lead to **runaway climate change.**

See Hyperfeedback

Climate Compatible Development

Development that is designed, built and managed to be both resilient and low or no carbon.

http://www.climateplanning.org

Climate Contests

One approach to encourage climate innovation is through the creation of contests.

Two of the best-known climate contests are the MIT Climate Colab annual contest and the Virgin Earth $25 Million contest to remove greenhouse gases from the atmosphere.

The Colab encourages collaboration among people from all over the world through its crowdsourcing platform. Over 500,000 have visited the Climate CoLab website, resulting in some 2,000 ideas submitted for judging and awards.

The Virgin Earth contest has selected 11 finalists, among them are companies focused on biochar, **holistic range management** and **direct air capture**. The award will be made when one of the finalists is determined to have met the selection criteria. Let's hope that occurs soon. Let's also hope that philanthropists, universities and governments establish more of these contests.

Perhaps the most valuable contest was announced by the MacArthur Foundation in 2016. Entitled 100&Change it will award $100 million for the proposal that has the best chance of bringing meaningful change in any field. While not specifically geared to climate change, the scale of this approach could serve as a model for a climate contest. Eight semi-finalists have been chosen with the winner to be selected in 2017.

See Climate Manhattan Project

climatecolab.org
https://www.macfound.org/programs/100change/
http://www.virginearth.com

Climate Debt

The climate debt movement argues that developed countries owe a debt to the developing countries as a result of the disproportionate costs of climate chaos that fall on these countries.

This movement, which emerged in 2009, divides climate debt into two components; adaptation debt and emissions debt, the first to compensate countries for the costs of adapting to the climate realities of the future and the second to compensate countries for agreeing to limit their emissions.

The US is estimated to owe developing nations some $4 trillion as its share of the total estimated climate debt of $10 trillion, according to a study published in *Nature Climate Chaos* in September 2015 by H. Damon Mathews.

The study does not suggest how the US and other developed nations will be required to pay this debt. It also does not quantify the cost to the developed world of unconstrained carbon emissions and minimal adaptation efforts on the part of developing nations. These costs include dramatic increases in migratory pressures, the collapse of economies and thus of import and export markets, increasing risk of climate induced wars.

$4 trillion paid out over 20 years ($200 billion a year or about 1% of GDP) would be a fair and affordable investment in discharging our estimated climate debt to underdeveloped countries while helping to enhance our economy and national security.

See Carbon Cost Shifting, Common but Differentiated Responsibilities. Contraction and Convergence

http://climate-justice.info/issues/climate-debt/

Climate Doomers

A number of scientists and activists believe that we have passed, or soon will pass, one or more climate tipping points, leading to the decimation of life on earth in the near future.

Among the most prominent is retired Professor Guy McPherson, who argues that the challenge now is to live without despair, and not focus on futile campaigns to reverse or stop climate chaos, since it's far too late. In a 2015 National Geographic documentary, hosted by the 'science guy' Bill Nye, McPherson predicted that human beings would be wiped out by 2030.

Another doomer is the British writer and activist Paul Kingsnorth. These and other doomers might also be called Collapseatarians or C and E'er's (Collapse **and** Extinction).

Many doomers urge a retreat to a simpler life, or as writer John Thackara says "our best course of action is to head for the hills with a truckload of guns and peanut butter."

Doomers can be distinguished from **eemergency** activists, who argue that the planet is on the edge of collapse but believe that a mobilization on the scale of WW2 to meet the **One Generation Challenge** can preserve at least some planetary habitability.

See Ameliorists, Brightsiders, Bright, Dark and Light Green Environmentalism, Climate Realists

http://news.nationalgeographic.com/2015/10/151131-climate-change-grief-bill-nye-explorer-television/

Climate Gender

A vastly unappreciated issue is the disparate ways in which gender affects and is affected by climate change.

Women in underdeveloped countries appear to suffer more from the effects of climate change, due to their lower social status, more limited physical survival skills, and social customs. Studies in Bangladesh, for example, show that women are more likely to be victims of cyclones than men.

Women in many countries tend to be more supportive of societal actions to address climate change than are men.

Surviving climate change without more closely focusing on the role of women would be a serious mistake.

Climate Gradualism

The belief that gradual actions to stem climate change are adequate or are all that is politically or economically feasible.

As we edge closer to **runaway climate change**, climate gradualism is likely to be seen as effectively equivalent to climate denial.

See Ameliorists, Brightsiders, Eemergency

Climate Hawks

Individuals who believe that aggressive action to address climate chaos is required. First formulated by Dave Roberts in Grist magazine in 2010 and derived from 'deficit hawks' — folks who urge aggressive action to reduce budget deficits. There are unlikely to be many who identify as both deficit and climate hawks.

Climate Inertia See Carbon Lock-in

Climate Insurance

Given the magnitude of climate-related projected losses and their unpredictability, it is unlikely that private insurance markets will be able to meet societal needs for insurance.

To help reduce the degree of climate related hardship and financial uncertainty, government subsidies and government sponsored insurance funds (such as the National Flood Insurance Program) will need to be rethought. New models of insurance to cover climate-related risks in less developed countries are also needed.

See Climate Equity Guarantee Agreements (CEGA)

Climate Positive Feedback

An effect such as the melting of sea ice that is enhanced or amplified by the consequences of the effect.

As when sea ice melts, the darker ocean water absorbs heat which results in higher temperatures, which leads to more sea ice melting and on and on until there is no more ice to melt.

A thermostat illustrates the opposite process of negative feedback. When the temperature goes above or below the setting the thermostat adjusts the temperature to return to the desired setting.

Unfortunately, climate chaos seems to be dominated by many more climate positive feedbacks (as many as 60 or more climate positive feedbacks have been identified) than climate negative feedbacks, strongly suggesting a grim prognosis for the habitability of the planet.

See Albedo, Hyperfeedback

Climate Proofing

Making a product, structure, community or home more resistant to the effects of climate chaos.

Climate Realists

Those individuals who by temperament, training and factual knowledge are able to assess the causes, consequences, and remedies for climate chaos without resort to denial or (excessive) alarmism.

Climate Rituals

Many are developing rituals to address climate chaos, as ritual helps people make sense of their lives and the world.

Perhaps the best known climate ritual is the Climate Ribbon. This international project uses art and ritual to provide an opportunity for people to grieve actual or potential losses and commit to taking action. One writes on a ribbon what he or she would most hate to lose to climate change and places the ribbon on a "ribbon tree" for all to see.

Expressing one's fears, griefs and hopes through art and ritual can be a profound opportunity for healing.

Ritual gets little attention from official bodies addressing climate chaos. This needs to change.

http://theclimateribbon.org/project#home

Climate Sensitivity

The relationship between the amount of CO_2 in the atmosphere and average global temperature increases.

Scientists estimate that a doubling of the CO_2 level from the preindustrial level of 275 to 550 parts per million would result in an increase of around 4°–6°. Well over half of that temperature increase comes not from CO_2 directly but from increased water vapor and decreased **albedo** due to sea ice melts. We are on target to reach this life-altering global temperature increase at about mid-century unless CO_2 emissions are immediately and drastically reduced.

Applying the common meaning of sensitivity, we can say with some confidence that most people (or at least their leaders) do not show high sensitivity towards the catastrophic consequences of a high degree of climate sensitivity.

https://www.theguardian.com/environment/climate-consensus-97-per-cent/2013/may/10/climate-change-warming-sensitivity

---

Climate Signals

These are "long-term trends and projections that carry the fingerprint of climate change," according to the website climatesignals.org.

ClimateSignals.org has identified at least 60 such signals including decreased surface wind speeds, drying out of soils and season creep.

These signals can be linked with individual events such as heat waves or floods and help us understand the relationship between greenhouse gases, signals and individual events.

Climate Stress Test

An analysis of the ability of an enterprise to handle the financial, emotional and physical consequences of climate chaos.

Modeled after bank stress tests created and imposed after the Great Recession (from the late 2000s until the early 2010s), these tests could be administered by governments, private organizations or self-administered.

Efforts by shareholders in 2016 to require climate stress tests at Exxon Mobil, Chevron and other companies lost by narrow margins in votes taken at their corporate annual meetings. One assumes that the reason the companies chose to fight these votes is that, had the resolutions passed, they would have been required to acknowledge the reality and magnitude of climate change, and its impact on their businesses.

And that wouldn't be good for business.

See Carbon Budget, Keep It in the Ground

http://www.wsj.com/articles/exxon-chevron-shareholders-narrowly-reject-climate-change-stress-tests-1464206192

Climate Test

A process proposed "to be used to evaluate all proposed energy supply and demand policies and projects in light of the globally agreed goal of limiting global warming to 1.5° C".

This definition is from the website climatetest.org, the organization or person (it isn't clear which) that developed the climate test website.

And with an estimated $6 trillion of infrastructure being built annually worldwide, a climate test is urgently needed to ensure that all infrastructure will be low or no carbon.

The concept of a test, standardized or not, to evaluate the consistency of any proposed project or policy with the Paris Climate Agreement is essential to operationalizing the agreement. A test also encourages greater awareness of the impact of investments on climate.

The federal government broadened its use of climate tests in rules adopted in 2016.

See Social Cost of Carbon

http://newclimateeconomy.report/2016/
https://insideclimatenews.org/news/03082016/obama-administration-climate-test-federal-projects-greenhouse-gases-emissions-keystone

Climate Triage

Difficult and contentious decisions will need to be made about which countries should receive international **readaptation** assistance to cope with climate chaos.

Should aid go to island nations whose disappearance is likely or to countries like nuclear-armed Pakistan where climate disturbances may bring catastrophe to the world? Or nations simply having the greatest per capita need — recognizing that need can be calculated in many ways.

Several ranking systems have been developed. The *2015 Notre Dame Global Adaptation Index* ranks New Zealand, Norway, Denmark, the UK and Germany as best equipped to deal with the pressures of climate chaos, while the Democratic Republic of the Congo, Central African Republic, Sudan, Eritrea, and Chad are the worst equipped.

The *2015 Maplecroft Climate Change Vulnerability Index* identified Bangladesh, Sierra Leone, South Sudan, Nigeria, Chad and Eritrea as the most vulnerable countries.

Ranking high on the vulnerability index does not mean that a country will necessarily benefit from climate triage decisions due to political, strategic or other considerations.

http://www.gain.org
https://maplecroft.com/portfolio/new-analysis/2014/10/29/climate-change-and-lack-food-security-multiply-risks-conflict-and-civil-unrest-32-countries-maplecroft/

Climate Tribes

In the US, climate chaos seems to have become a litmus test that identifies one as belonging to one of two antagonistic tribes. When we argue about climate change Yale Professor Dan Kahan says, "What we're actually arguing about is who we are, what our crowd is, who our tribe is."

The tribal model appears to explain the extreme polarization of the population on a host of other issues including electoral behavior.

It also suggests that more education, facts and reason won't help to convert people who are climate deniers, according to Kahan.

Studies show that where facts have assumed symbolic significance for individuals with opposing cultural identities, trying to correct mistaken beliefs with "corrective" information tends to backfire and harden the message recipients' resolve.

See Backfire Effect, Denial Triad, Social Proof

https://blogs.scientificamerican.com/guest-blog/the-meaning-of-scientific-truth-in-the-presidential-election/

Climate Victory Bonds

Small denomination bonds citizens can buy to finance local or national climate mitigation and adaptation investments.

Legislation has been proposed in Congress to facilitate the issuance of these bonds, but has received no support from the climate denial community. Big surprise!

Coal as a Weapon of Mass Destruction

Each plant consumes a considerable amount of the world's remaining **carbon budget**, ensuring that our **post normal climate** will be **climatastrophic** for most life on Earth.

What kind of species, armed with the incontrovertible knowledge that carbon emissions irrevocably changes our climate, continues to operate scores of facilities that are guaranteed to doom the planet? And this destruction is not in some faraway place a long time from now, but in every corner of the planet in this century. Virtually every person involved in the design, construction and operation of these plants will live long enough to experience the terrible consequences of what they are doing.

We're playing planetary Russian roulette with a loaded gun.

http://newclimateeconomy.report/2016/

Collective Empathy

The emotional connection that individuals or groups of people have for other groups.

This quality becomes essential to cultivate as climate chaos requires community cohesiveness and international solidarity for our survival. The cultivation of collective empathy could be thought of as a central goal of human experience, albeit one that continues to elude us. Some research has shown a direct link between those most prone to climate denial and lack of empathy.

See Climadelic Therapy

http://uu.diva-portal.org/smash/record.jsf?pid=diva2%3A945529&dswid=-9866

Collective Inaction See Social Proof

Common but Differentiated Responsibilities

These four words (Officially it's 12 words and one helluva of an acronym *CBDRILONCWRC* ("*Common but Differentiated Responsibility in Light of National Circumstances with Respective Capability*") encapsulate the relationship of developed and less developed countries with respect to their responsibilities to reduce carbon emissions.

This formulation is an artful attempt to bridge the profound and historic gap between those countries that have benefited from high levels of carbon emissions in the service of building advanced societies over the past 150 years or so, and those (generally the less developed countries) that did not have the opportunity or good fortune to develop prior to the time when carbon was identified as the primary culprit in climate destabilization. These less developed countries will generally suffer more from the consequences of the **post normal climate** and have fewer resources to protect their people and national wealth.

The four words mean that while we're all in this together (common) some countries have greater (differentiated) responsibilities than others to address the causes and consequences of climate chaos.

This principle has been a bedrock of climate discussions since 1992, and has been interpreted in different ways over the years.

See Carbon Cost Shifting, Climate Debt, Contraction and Convergence

http://www.climatenexus.org/about-us/negotiation-issues/common-differentiated-responsibilities-and-respective-capabilities-cbdr

Consumption-based Footprint

The measurement of the CO_2 generated by an individual or group based on where the products are consumed rather than on where they are produced.

Consumption-based footprinting increases the size of the footprint of countries that import carbon intensive products while reducing the footprint for those countries exporting large quantities of carbon producing products such as China.

In the United Kingdom, for example, consumption-based footprinting contradicts the widespread belief that carbon footprints within the UK have declined significantly in recent years. Consumption-based footprinting shows that emissions declined just 7% from 1992 to 2012 versus the 25% that the standard territorial-based footprint shows.

http://www.emissions.leeds.ac.uk/

Contraction and Convergence

The principle that a worldwide contraction in the volume of carbon emissions should occur along with a reduction in the disparity between poor and wealthy nations in per capita carbon emissions.

While most countries agree with this basic principle, conflicts arise over the rate of contraction and the degree of convergence.

For example, a contraction scenario that calls for zero net emissions worldwide by 2050 suggests an annual reduction of CO_2 of 2% to 3%, whereas reaching zero emissions by 2030 would require a reduction rate in excess of 6% per year, a rate never seen in modern history.

Convergence, based upon the principles of fairness and equity, requires wealthier countries to decarbonize at a much faster rate than less wealthy countries. The poorest countries are encouraged to increase their per capita carbon consumption for a time to provide basic electricity and other services for their populations.

The United States burns about 16 tons of CO_2 a year per capita, the European Union burns seven tons and most African countries burn one ton or less. Under a contraction and convergence policy by say 2025 the US might burn 13 tons of CO_2 a year, the EU five tons a year and African countries two tons a year.

See Common but Differentiated Responsibilities, Decarbonization Divide

http://blogs.law.widener.edu/climate/2014/02/21/10-reasons-why-contraction-and-convergence-is-still-the-most-preferrable-equity-framework-for-allocating-national-ghg-targets/

COP 21 (Conference of the Parties 21st Meeting)

The 21st meeting of the Conference of the Parties took place in Paris in December 2015 as part of the "The United Nations' Framework Convention on Climate Change" (UNFCCC).

Bringing together 195 countries over two weeks COP21 achieved a historic agreement to tackle climate change. While many hailed the agreement as a breakthrough, others criticized its deficiencies and found it to be wholly inadequate to head off catastrophic climate chaos.

The most important outcome is not found in the specific provisions of the agreement. Rather, it is the unambiguous message that the world is serious and united in the necessity of fighting climate change and getting off fossil fuels. (Though the term 'fossil fuel' is not mentioned in the agreement, supposedly because of objections from Saudi Arabia.)

Whatever the merits of the agreement, one must be impressed with an outcome that received the support of many major environmental groups and several major oil companies.

Some key provisions of the agreement:

Voluntary agreement with no sanctions imposed for missing goals

Goal is to keep temperature increases well below 3.6º and to pursue efforts to limit temperature increases to 2.7º

Five year reviews of progress starting in 2018, with the agreement going into effect in 2020 and expiring in 2030

No mention of aviation or shipping (A modest aviation agreement was reached in 2016)

Requires countries emitting 55 percent of emissions and 55% of the population to ratify before it goes into effect with the deadline being April 21, 2017. (Agreement was ratified in October 2016)

Sets a minimum of $100 billion per year in aid to developing world

Creates a process to examine 'loss and damage' issues regarding compensation to less developed countries for the climate costs the developed countries have imposed on them, but specifically rules out payments for damages

Achieving each country's voluntary climate goals as set forth in the **INDC's** results in projected temperature increases well above the 3.6º target, according to several analyses performed by third parties.

Did we witness 'climate peace in our time' in Paris in December 2015? Will Paris achieve anything more than British Prime Minister Neville Chamberlain achieved in 1939? We may not know for some time.

The November 2016 COP 22. held in Marrakesh, Morocco focused on implementation of the Paris Climate Agreement. The United States began the process of withdrawing from the agreement in 2017.

See Climate Debt, Contraction and Convergence, Decarbonization Divide

Coral Bleaching

Coral expels algae when stressed. This lightens the color of coral and often leads to coral death.

Coral bleaching is now widespread. As some 25% of all marine life avail themselves of the food, shelter and other benefits of coral reefs, the consequences of bleaching are devastating for the vitality and biomass of the oceans, and ultimately for the ability of people to obtain nutrients from the ocean. Many scientists believe coral bleaching is a key **climate canary**.

While most coral bleaching is related to climate chaos, infections, pollution, cold water and even sun-screen lotion can lead to bleaching.

While there are no foolproof ways to prevent climate chaos linked bleaching or to reverse bleaching that has occurred, creating marine reserves that limit pollution and fishing appears to help restore reefs.

See Marine Heat Waves, Ocean Hawks, Reef Races

Dark Snow

Deposits of soot from industrial processes, diesel engine emissions, dust storms and fires, as well as from ground bare spots due to snow melting, has led to a noticeable darkening of snow in many regions of the world, including the Arctic and the Himalayas.

This darkening of snow reduces the **albedo**, and thus accelerates the heating of the atmosphere. This is one of many **climate positive feedbacks** accelerating the pace of climate chaos.

Many scientists believe that focusing on reducing soot may be one of the most effective ways to reduce the increase in global temperatures.

See Socialized Soot, Watermelon Snow

https://www.theguardian.com/environment/2016/dec/21/cutting-soot-emissions-arctic-ice-melt-climate-change?CMP=Share_iOSApp_Other
darksnow.org

De-extinction See Civilizational Preservation Movement

Degrowth

A social movement originating in Europe whose goal is to replace the mainstream model that economic growth as measured by GDP is always desirable. The degrowth movement believes that ecological and quality-of-life goals can best be realized by shrinking quantitative growth, while simultaneously enhancing quality of life.

Most degrowthers would likely agree with the writer and **planetarian** Paul Hawken:

> At present, we are stealing the future, selling it in the present, and calling it gross domestic product.

Degrowth is seen by its advocates as the only way to attain safe levels of CO_2 in the atmosphere. Other environmental benefits include reducing the rate of species and habitat loss, biodiversity preservation and a reduction in air and water pollution.

It may well be that degrowth will occur, either voluntarily after a lot of political and economic chaos or involuntarily after a lot of climate chaos. Which would you choose?

See Bright, Dark and Light Green Environmentalism, Decarbonization Divide, Smart Degrowth, Syntheticism

http://clubfordegrowth.org

Direct Air Capture

One of the emerging technologies for removing CO_2 from the atmosphere.

Direct air technologies capture CO_2 directly from the atmosphere as opposed to removing it from specific sources such as power plants and factories. This approach may be easier to scale up with some rapidity, unlike most proposed **negative emission** technologies.

Direct air capture is likely to require lots of energy (resulting in a low **EROEI**) as it has to overcome the inherent inefficiency of finding and extracting a substance (CO_2) that is present in very low concentrations in the atmosphere.

Several companies are racing to develop direct air capture systems.

See Climate Solution Contests, Negative Emissions

Drunken Trees

The thawing of permafrost creates uneven and unstable topography leading to trees leaning and even toppling throughout Arctic latitudes, much like a drunk might look or act.

And as with inebriated people, collateral damage occurs: homes, roads and other infrastructure are damaged or destroyed. Unfortunately, drunken trees are unlikely to be helped by 12-step programs. Only when humans stop drinking the carbon nectar that powers our civilization will drunken trees have a chance to recover.

See Carbon Sirens, Permadeath

Ecocide

The destruction of the environment as a result of intended or unintended consequences of human actions.

A political/legal movement has emerged in Europe that would make ecocide a crime and require nations to intervene to protect natural resources when at risk. An International Environmental Court would be created to adjudicate these claims and issues.

See Carbon Crimes Against Humanity, Children's Climate Crusade

eradicatingecocide.com

Embodied Carbon/Energy

Embodied energy/carbon is the amount of carbon or energy used to fabricate a material or product through all stages of the extraction and transportation process.

Embodied energy and carbon is a topic of rising importance as governments and the private sector search for cost effective ways to reduce carbon emissions.

Empathy Deficit Disorder

Empathy deficit disorder describes the all-too-common lack of empathy in our society.

While some would argue that empathy is not in deficit, how would they explain the callousness of the developed world towards assisting less developed nations in both mitigating and adapting to climate chaos?

As to why empathy appears to be in short supply when the demand is so high, at least part of the answer lies in the disconnect with nature that many in the affluent world experience. It is difficult to be fully *in* nature without feeling a degree of awe, gratitude and connection with the wonders and mysteries of life. These qualities correlate highly with empathy. Climate denial is also associated with a lack of empathy.

The absence of empathy and a connection to nature is linked to another quality of mind — reductionist one dimensional thinking.

A 2016 essay by Charles Eisenstein identified the pervasiveness of reductionist thinking as a major contributor to the world's problems, including climate chaos. As he says:

> This quality of complex systems collides with our culture's general approach to problem-solving, which is first to identify the cause, the culprit, the germ, the pest, the bad guy, the disease, the wrong idea, or the bad personal quality, and second to dominate, defeat, or destroy that culprit. Problem: crime; solution: lock up the criminals. Problem: terrorist acts; solution: kill the terrorists. Problem: immigration; solution: keep out the immigrants. Problem: Lyme disease; solution: identify the pathogen and find a way to kill it. Problem: racism; solution: shame the racists and illegalize racist acts. Problem: ignorance; solution: education. Problem: gun violence; solution: control guns. Problem: climate chaos; solution: reduce carbon emissions. Problem: obesity; solution: reduce caloric intake.

See Climadelic Therapy, Collective Empathy

Energy Density/Sprawl

The amount of energy contained or available per unit area. Fossil fuels are substantially more energy dense than renewables, but not nearly as energy dense as nuclear energy.

Most of us rarely think of the energy miracle that an energy dense fuel like oil is. For example, a gallon of gasoline is equivalent to the energy that a person doing manual labor would burn off in some 80 days. The energy density of a fuel is an extremely important consideration as lower energy density requires either more land or other resources to convert it into useful energy.

Energy Return on Energy Invested (EROEI)

The ratio of the amount of energy produced versus the energy required to produce it. This ratio has steadily declined for fossil fuels as extraction has become much more resource intensive given the remoteness of the remaining oil fields, and their limited accessibility.

Vigorous debate surrounds the significance of the long-term decline in the EROEI for oil and gas. Some have argued that the ratio is becoming so low that oil companies (or even, some say, all of civilization) will decline or collapse given the central role that a high EROEI has played in creating the modern high energy high consumption world.

As the well-known energy analyst and author Richard Heinberg has said:

> As much as I hate to think so, thermodynamic decline (e.g., a low EROEI) and economic contraction could seriously impair our chances for a robust renewable energy transition in response to the threat of climate change. Building enough solar panels and wind turbines, and adapting the ways we use energy (in building heating, in industrial processes, in transportation, in food systems, and on and on), will take time and many trillions of dollars of investment. It will also require stable international markets and supply chains, and those could be thrown into turmoil by the declining thermodynamic profitability of our society's current primary energy source — unless we can somehow build a bridge to the future while the highway we're on is crumbling beneath us.

Event Compression

The frequency with which infrequent climate related events occurs is increasing. A flood that occurs once in 100 years now is occurring much more frequently in many locations. In the United States alone it is estimated that eight 500-year storms occurred during the first eight months of 2016.

After decades of relative climate stability, predicting the frequency of various weather events such as floods and hurricanes will be exceedingly difficult. Location and relocation decisions, insurance coverage and premiums, and decisions on rebuilding homes and public facilities will have to be made in an environment of much uncertainty.

See Ecostasis, Never Normal Coalition, Predictable Unpredictability

Fashionable Pollution

The clothing industry is the second largest CO_2 polluting industry in the world. The average person worldwide buys 20 changes of clothes each year. Perhaps we can do with 10 (or even five) and help preserve life on earth?

http://www.esquire.com/style/news/a50655/fast-fashion-environment/

Fee and Dividend

A proposal to reduce carbon emissions that places a progressively higher fee or tax on carbon over time.

The fees are levied at the wellhead, mine or entry point into the country to simplify administration and compliance. Countries exporting products to the US will be subject to a tariff unless the country has levied a comparable carbon fee.

The revenues these fees generate are returned as dividends to all residents of the jurisdiction on an equal basis. Most residents will enjoy a windfall; generally, only the high carbon generating higher income residents will pay more in higher prices than they receive in dividends.

The goal of the fee and dividend approach is simple. Reduce carbon emissions through a simple and fair mechanism that benefits those who use the least carbon in their lives.

Prominently supported by the Citizens Climate Lobby and the Climate Leadership Council, this effort, while viewed favorably by many economists and climate experts, has been unable to overcome objectives by Republican legislators and interest groups. Alternative approaches include **Cap and Trade, TEQs, Carbon Taxes,** Cap and Dividend, rationing and exhortation.

https://citizensclimatelobby.org/basics-carbon-fee-dividend/

4/1000 Initiative

There's nothing under the ground that's worth more than the little layer of topsoil sitting on top of it.

Wendell Berry, Farmer and Author

4/1000 is an international effort unveiled at the COP21 conference in Paris in 2015 to encourage soil management practices that could lead to an increase in the carbon content of the soil by 4 parts per 1000 annually. The French government along with 37 other governments and the many non-profits that organized this effort estimate this initiative has the potential to offset as much as 40% of human generated carbon emissions.

The practices to be encouraged include: cover cropping the soil, restoring degraded lands, planting legumes and crops that restore nitrogen, conserving water and nourishing soils with manure and compost.

These practices will also improve the health of the soil, increase plant yields and reduce runoff and land pollution. Increasing the health of the soil is critical, as the UN Food and Agricultural Organization has estimated that all the top soil in the world could be gone in 60 years. Yes, let me repeat that. All the topsoil in the world gone in 60 years with agricultural business as usual.

Successfully implementing this program at the necessary scale will be quite challenging. Bringing these practices to lands around the world requires thousands of skilled and experienced agricultural agents knowledgeable about local conditions. Opposition from agricultural companies and local farmers will be encountered. Climactic disruptions will present a real risk to the program. Reliably measuring whether soils are absorbing more carbon is also a challenge.

See Agroecology, Plant Purgatory, Savory Holistic Range Management, Terra Preta

http://4p1000.org/understand
https://www.scientificamerican.com/article/only-60-years-of-farming-left-if-soil-degradation-continues/

Fugitive Emissions

Some CO_2 emissions result from leakages, particularly leakages coming from the exploration, distribution and combustion process for natural gas. These leakages are known as fugitive emissions.

The degree to which natural gas generates fugitive emissions is a contentious issue, with studies supporting each side in this debate. If fugitive emissions offset the lesser carbon emitted by natural gas as compared to coal, then the so-called bridge to the renewable future promised by gas may be in danger of falling victim to a carbon earthquake.

The extent of fracking, the future of coal, the timeline for bringing renewables online, and the pace of global heating all depend in part on the degree to which fugitive emissions are generated by natural gas, and whether better natural gas management can reduce these leaks.

Geoengineering

A conscious effort to manipulate the climate on a large-scale basis with the goal of stopping or slowing climate change.

Geoengineering is perhaps the most contentious of all approaches to CO_2 emission control. Many geoengineering ideas have been proposed, including spraying aerosols or erecting mirrors in the atmosphere, and adding iron to the seas.

The difficulty in accurately forecasting the consequences of a given technique, the likelihood of severe weather changes unequally distributed throughout the world, the challenge of agreeing on a process for deployment, and the likelihood of some significant degree of **moral license** leading to greater carbon emissions are among the extremely difficult issues that are nowhere near to being resolved.

An example of the problematic yet enticing nature of geoengineering is shown by a study done in September 2016 by the National Center for Atmospheric Research. Injecting sulfates into the atmosphere — essentially mimicking what happens with volcanic reactions — would require 18 gigatons or 39,600,000,000,000 (!) pounds of sulfates a year for 160 years to potentially limit temperatures to a 3.6° increase rather than a 5° or larger increase. That effort would do nothing for ocean acidification, could harm the ozone layer, and would have uncertain but highly significant impacts on temperatures and rainfall throughout the world.

The only way to avoid serious consideration of geoengineering is the immediate transformation of every major country's economy through implementation of the **One Generation Challenge.** Yet this effort to reach zero carbon has little support — for now.

See Carbon Maze, Termination Shock

https://www.scientificamerican.com/article/geoengineering-and-climate-change/
http://www2.ucar.edu/atmosnews/just-published/122687/2-degree-goal-and-question-geoengineering
http://www.newyorker.com/magazine/2012/05/14/the-climate-fixers

---

Geophony

Naturally occurring sound generated by inanimate objects.

These sounds include water running, rocks falling, wind rustling, earthquakes, ice quakes, volcanos, thunder and rainfall. Geophony is changing due to climate chaos with unknown consequences for life.

The concept was developed by the soundscaper Bernie Krause.

See Biophony

Global Dimming

Global Dimming results from a decrease in the amount of solar radiation reaching the surface of the earth as a by-product of fossil fuel related pollution.

Tiny pollutant particles absorb solar energy and reflect sunlight back into space.

The decreasing solar radiation offsets a significant portion of carbon emissions. Sharp CO_2 reductions will reduce global dimming/ aerosol cooling and will increase global temperatures, thus paradoxically worsening climate chaos. If all carbon emissions were stopped tomorrow, scientists calculate that the earth's warming could perhaps increase by about 1° in the short run.

No widely-accepted strategies to counteract the expected increase in atmospheric temperatures from a reduction in **global dimming** exist.

See Carbon Maze

http://www.conserve-energy-future.com/causes-and-effects-of-global-dimming.php

Global Warming

The increase in global temperature resulting from adding one or more greenhouse gases to the atmosphere.

The term was first used in a Science article in 1975 by geochemist Wallace Broecker of Columbia University. Prior to that time there was less certainty about the direction of temperature changes with some scientists believing that global cooling was possible. The term 'inadvertent climate modification' was therefore used until it became clear that warming was much more likely to occur than cooling.

See Climate Change, Global Weirding

Global Weirding

Clear air turbulence, dark snow, ice quakes, marine heat waves, sunny day flooding, drunken trees, jelly lands, methane craters, watermelon snow, zombie bacteria, and even a parking garage octopus are a few of the spookier **post normal climate** changes occurring.

There should no longer be any doubt about the aptness of the term global weirding that many are using to describe climate change.

Green Assets

Assets that contribute to the functioning of a **zero-carbon** economy.

Examples include wind turbines, solar thermal installations, solar panel factories, **CCS** and **BECCS** plants, pedestrian and bicycle paths, walkways, electric cars and car charging stations. In addition, green infrastructure such as parks, gardens, forests, areas of **blue carbon** and natural stormwater and wastewater systems are green assets.

Green assets are contrasted with stranded assets.

See Active Infrastructure, Stranded Studies

Green Banks

Publicly owned banks that provide loans and technical support for renewable energy and other climate appropriate investments. Five states — Connecticut, California, Hawaii, Rhode Island and New York — and one county, Montgomery Co, MD, have set up green banks as of 2016.

Green Bonds

Bonds that are intended to be invested in projects that reduce climate emissions or help communities adapt to climate chaos.

Green bonds are issued by many organizations and governments. These bonds are generally not regulated and are only sometimes certified by third parties. Over $81 billion of these bonds were issued in 2016.

Green Brain

A term encompassing the psychology and behavioral economics of climate communication, formulated by Simran Sethi, a sustainability educator and author.

Greenhouse Gas

Any atmospheric gas that traps heat leading to an increase in atmospheric temperatures. The three most important greenhouse gases are carbon dioxide, methane, and nitrous oxide. Other greenhouse gases include water vapor, chlorofluorocarbons and hydrofluorocarbons.

See CO_2e

Greenhouse Gas Intensity

The ratio of greenhouse gas emissions to total energy consumption.

The George W. Bush administration focused on reducing greenhouse gas intensity rather than reducing **greenhouse gases**. The problem with greenhouse gas intensity is that improvements in efficiency and changes in the mixture of goods and services consumed could lead to a reduction in greenhouse gas intensity while still resulting in an increase in greenhouse gas production.

The earth is indifferent to greenhouse gas intensity. All it responds to is the atmospheric concentration of greenhouse gases.

Green Tea Party

A faction of the Tea Party that has cooperated with environmental groups to support putting clean energy and solar power on equal footing with other forms of energy. This movement emerged from political battles in several southern states when there was an attempt to put restrictions on the use of decentralized solar power. The green tea party helped defeat a Florida ballot initiative in 2016 that would have made solar panels more difficult for homeowners to get.

Green Tea Party supporters mostly remain in the climate change denial camp and support solar power for free market and not environmental reasons.

http://www.pri.org/stories/2015-04-11/green-tea-party-fights-more-environmentally-friendly-gop

---

Bates/Heisenberg Climate Uncertainty Principle

One of the challenges of fighting climate chaos is having some sense of its pace and intensity. Longtime climate activist Albert Bates proposed an analogy with the Heisenberg Uncertainly Principle. This well-known principle of physics suggests that one can't know both the exact location and the exact speed or momentum of a particle, due to the wave-like nature of matter. Bates argues that similarly one can't know both the timing and the certainty or intensity of climate change.

An interesting and provocative, though not particularly cheerful, idea.

Hibernation Hiatus

Some animals are seeing their hibernations become disrupted as a result of climate chaos driven changes in seasonal temperatures.

European brown bears in Spain no longer hibernate, young hedgehogs in England are starving due to later births, and many marmots in the Rocky Mountains are starving because of an early awakening from hibernation when snow still covers the ground, thwarting their ability to find food.

You can be sure that these aren't the only animals to be stressed or even killed by **climacide**.

See Biotic Migration

Hydropolis

A city built on or around water.

Floating cities may be able to withstand some climate shocks better than the standard fixed location land-based city. Look for the creation of floating cities or at least floating neighborhoods.

http://www.theverge.com/2016/10/27/13418576/arx-pax-floating-cities-climate-change-hendo-hoverboard

Ice quake

Ice earthquakes result from rapid warming and cooling of ice covered terrain.

Ice quakes are increasing in frequency and intensity as a result of climate change; also known as cryoseisms.

Intended Nationally Determined Contributions (INDC)

In typical UN style '*Intended Nationally Determined Contributions*' is the name used to describe the plans that each nation prepares that indicate how the country is proposing to reduce greenhouse gas emissions between 2020 and 2030.

How much more understandable and accessible would these plans be if they were called something simple like '*climate change action plans*'?

These plans were prepared just prior to the 2015 Paris COP 21.

They vary widely in rigor and detail as well as in their climate ambitions. A recent analysis of INDCs from African countries included this comment from a knowledgeable observer:

> The process through which many of these INDCs were written was seriously fraught with error, with minimal stakeholder consultation, data gathering and analysis... It is understandable that France wanted to get an agreement, but I fear the success in Paris may have come at the expense of African countries.

Achieving the voluntary goals outlined in each national plan would result in substantial increases in worldwide average temperatures to well above the 3.6º international target. Studies estimate that temperatures could rise anywhere from 4° to 6° even with full implementation of each country's plan.

These plans will be revisited in 2018.

See COP21

http://www.climatechangenews.com/2016/11/03/africas-buyers-remorse-over-Paris-climate-deal/

Intergovernmental Panel on Climate Change (IPCC)

The IPCC is a non-governmental body established by the United Nations in 1988 to produce reports synthesizing existing climate related research.

The IPCC is considered as the most authoritative source of climate change related data and forecasts. The IPCC has issued 6 reports, each including a consensus-based summary report that requires word-by-word vetting by member countries. Given the complex and lengthy report preparation process, it is no surprise that the analysis and conclusions reached are increasingly seen as seriously out of date by the time of publication. The next major IPCC report is expected in 2022.

See COP21

Intermittency

The uneven intensity of wind and solar energy in a given location over time.

Wind blows and the sun shines periodically and with much less than 100% predictability. This is a major economic and logistical problem for utilities as they need to have reliable and sufficient dispatchable (turned on quickly) power. As the proportion of wind and solar power used for electricity increases intermittency becomes an even greater problem.

Many storage approaches are being developed, tested and deployed to address intermittency. While much progress has been made, as yet there is no widely-accepted and available remedy to fully overcome the problem of intermittency.

See Virtual Power Plant

Internal Carbon Price

The establishment of a benchmark price for carbon emissions within an organization.

An internal carbon price evaluates investments and other expenditures to determine their impact on carbon emissions with the goal of reductions in these emissions. Hundreds of companies are now using internal carbon prices, many in anticipation of carbon caps or taxes.

See Social Cost of Carbon

Kayactivist

People who have paddled kayaks to block or otherwise disrupt oil carriers or related vessels.

Kayactivists have had considerable success in thwarting BP's effort to sail from Seattle to explore potential oil fields near the Arctic Circle.

Kaya Identity

A formula developed by Yoichi Kaya and colleagues at Tokyo University to calculate the total CO_2 emissions generated by a region or country.

The formula determines total CO_2 emissions by calculating the product of population, GDP per capita, energy use per unit of GDP, and carbon emissions per unit of energy consumed. As the graph below shows, significant increases in CO_2 have occurred since 1970 despite improvements in energy intensity and the carbon content of energy. Population growth and income growth per capita explain the increases in CO_2.

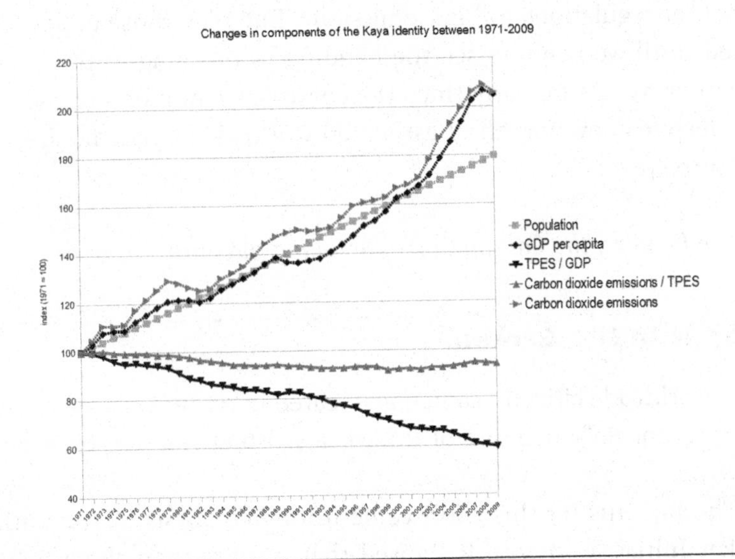

Keeling Curve

A graph displaying the rising concentrations of CO_2 in the atmosphere.

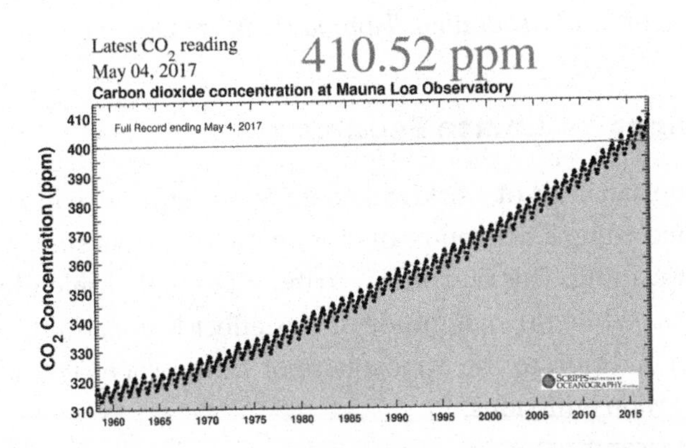

The methodology was first developed by David Keeling in 1958 using a device that collected data at the Mauna Loa Observatory in Hawaii. This graph is one of the more important scientific measurements in history, as perhaps the most critical number to track climate change is the concentration of CO_2 in the atmosphere.

So yes, let's celebrate when solar panels or wind turbines are deployed. Let's also celebrate when communities adopt carbon taxes and other regulations to limit emissions. But keep those celebrations muted until we begin to see the bending or breaking of the Keeling carbon curve. Particularly since the curve has continued to rise each year despite a leveling off of estimated worldwide CO_2 emissions the past three years.

See Bending the Carbon Curve, Breaking the Carbon Curve

Keep It in the Ground

A worldwide citizens' campaign created in recent years whose goal is to prevent 80% or more of known fossil fuel reserves from being burned.

The impetus for this effort came from an analysis by the Carbon Tracker Initiative in 2011. It showed that if more than about 20% of known fossil fuel reserves are burned the world would exceed the 3.6º limit generally recognized as the most that temperatures can increase without causing catastrophic global scorching.

See Intentional Grounding, Tipping Point, 2° Dogma

Land Carbon-Climate Feedback

A recent analysis of worldwide data by 49 authors concludes that soils are becoming a net *source* of carbon rather than a *sink*, as a result of global warming. This reversal is attributed to higher microbial respiration rates. While the magnitude of this effect is uncertain, estimates suggest an addition to the atmosphere of from 12% to 17% of human induced carbon emissions.

This presents yet more compelling evidence that an **eemergency** WW2-scale program is necessary to have any reasonable chance to escape **climatastrophe**.

See Carbon Maze, Climate Positive Feedback

http://www.nature.com/nature/journal/v540/n7631/full/nature20150.html

Living Fences

The use of trees and shrubs as fences is one of many modifications to conventional agricultural practices that can reduce carbon emissions, provide environmental benefits and enhance aesthetics.

Living Shorelines

Shoreline protection that relies on natural features including greenery and berms rather than carbon intensive concrete or other industrial structures for flood protection.

Living shorelines are an important component of community resilience. They can minimize storm damage, improve soil quality and provide wildlife feeding areas. And they're much nicer to look at — at least to most of us — than concrete. (A letter from the concrete industry is expected.)

Long Zoom

A term used by the writer Steven Johnson that is "the attempt to explain historical change by simultaneously examining multiple scales of experience."

Climate change is a prime example of a phenomena where a long zoom approach is essential to understanding the complexity of the issue.

As the breadth of the *CVF* illustrates, climate understanding requires knowing some atmospheric physics, abnormal psychology, marine biology, disaster management, insurance markets and a hundred other disciplines. And most importantly a knowledge of the connections between them.

We urgently need to find ways to inculcate a collective long zoom sensibility to help master life in a **post normal climate**.

See Climate Universities, Misattribution Bias

Man-Made Natural Disasters

Natural disasters that are either largely caused or made more severe by climate related human influenced events including fires and floods.

A **post normal climate** will result in many more man-made natural disasters. President Obama has spoken of man-made natural disasters in remarks on climate change.

See Disaster Emissions

Marine Heat Wave

This higher than normal warming of large areas of the ocean was first identified in 2011.

A 2015 marine heat wave near Australia killed off large areas of kelp and mangrove forests with water temperatures as much as 8° *higher* than surrounding areas for many weeks. A similar marine heat wave affected the Pacific Ocean off the US west coast and led to temperature increases as high as 7° and the widespread death of marine life. It should be noted that coral reefs can begin to bleach when water temperatures increase by as little as 1.6° for a few weeks.

Scientists believe that these marine heat waves are becoming more intense and frequent due to climate chaos.

See Global Weirding, Ocean Hawk

http://www.nationalgeographic.com/magazine/2016/09/warm-water-pacific-coast-algae-nino/

Marsitect

An architect who designs structures for living on Mars.

Why is this entry here? Because the desire (and soon perhaps desperation) by many to escape Earth to Mars or other planets will grow.

Vera Mulyani founded Mars City Design to design a community for 1,000 Earthians. Isn't it good to know that there's at least one Marsitect designing your new residence?

See Help Kelp, Reef Races

Methane Craters

Craters caused by methane explosions have been discovered in Siberia since 2013.

Several areas of land with jelly-like consistency have also been discovered. CO_2 and methane levels at these **jellylands** are as much as 20 times higher than the norm for CO_2 and 200 times higher for methane. As permafrost begins to thaw and CO_2 and methane escape, dangerous increases in global temperatures are likely.

See Climate Positive Feedbacks, Drunken Trees, Global Weirding, Permadeath, Tipping Points, Zombie Bacteria

http://news.nationalgeographic.com/news/2015/02/150227-siberia-mystery-holes-craters-pingos-methane-hydrates-science/

Methane Hungry Bacteria (Methanotrophs)

Methanotrophs have the promising ability to absorb methane.

They are potentially one of the all-too-rare negative climate feedbacks, as they increase their absorption of methane as temperatures heat up, according to a 2015 study by Princeton University. To what degree this could offset the positive **soil carbon-climate feedback** recently identified is a critical question.

https://blogs.princeton.edu/research/2015/08/14/on-warmer-earth-most-of-arctic-may-remove-not-add-methane-isme-journal/

Misattribution Bias

Assigning responsibility or cause to the wrong event or phenomena as a result of looking only at superficial or first order causes.

An example is attributing the chaos in Syria to political or factional factors and overlooking or rejecting the role of climate chaos related drought on the origins of the war.

Severe winter snowstorms are seen by some as a refutation of climate change. Yet by increasing water vapor in the atmosphere, global heating can lead to greater storm severity in winter as well as summer.

Identifying the wrong cause of events leads to wrong actions taken in response to these events.

Moral License

Feeling virtuous after engaging in an altruistic or unselfish behavior often leads to a conscious or unconscious belief that it's fine to offset that behavior.

An example of this well-known phenomenon is to outfit your house with LED lightbulbs, buy a hybrid car and then feel entitled to fly to Europe. This has been characterized by Tim Berners-Lee and Duncan Clark in their book *The Burning Question* as a "kind of ethical rebound effect". Moral license is an example of yet another barrier encountered while navigating the **carbon maze**.

See Backfire, Carbon Compensatory Behavior, Rebound Effect

Negative Emissions

A process, condition or technology that results in more carbon removed from the atmosphere than enters it, thus reducing carbon concentrations.

Negative emission technologies include **BECCS**, **direct air capture** as well as natural processes and technologies such as soil carbon sequestration, restoration of wetlands and enhanced rock weathering.

The Paris Climate Agreement relies on negative emissions to bring carbon concentrations and temperature levels back down after 2050. While implementing negative emissions presents many challenges including cost, political acceptance and technological feasibility, perhaps the biggest challenge is the ability to scale up to the extraordinary levels needed to make a significant dent in atmospheric carbon concentrations. One study estimated that the land area needed for BECCS to replace fossil fuel emissions could be larger than that of the US.

See Carbon Drawdown, 4/1000 Initiative, Terra Preta, Savory Holistic Range Management

https://www.carbonbrief.org/explainer-10-ways-negative-emissions-could-slow-climate-change
https://www.ncbi.nlm.nih.gov/pmc/articles/PMC2859252/

Nemesis Effect

In 2000, Chris Bright of the Worldwatch Institute introduced the term "nemesis effect" which refers to the cumulative effect of multiple stressors and conditions that lead to unanticipated consequences. Which sounds like a perfect description of climate chaos.

See Hyperfeedback

Neoskeptics

A term coined by Paul Stern of the National Research Council that describes individuals who accept the reality of climate chaos, but who do not believe in taking any positive action to fight it.

The refusal to take action often results from skepticism that addressing its causes and consequences will have any noticeable impact, or is too costly. Neoskeptics appear to be what might be called 'soft climate deniers'.

See Ameliorists, Brightsiders, Climate Doomers, Denial Triad, Denial Two Step

---

NoUMBY (Nothing Under My Backyard)

The rallying cry of those seeking to ban fracking now and the cry of those who may potentially seek to ban sequestered carbon under their backyards in the future.

Oh Shit Moment

That moment when a person first recognizes in his or her gut the immensity of the reality, immediacy, and horror of climate chaos, and the extraordinary difficulty of combating it. This sinking moment is one that may call for support from friends or family to help process the feelings of helplessness and unreality often associated with one's oh shit moment.

First described by the author Mark Hertsgaard in 2009.

See Climadelic Therapy, Climate Fatalism

http://markhertsgaard.com/climate-roulette/

Participatory Climate Budgeting (PCB)

Over 1500 jurisdictions and agencies throughout the world have empowered their citizens or stakeholders to determine some part of their budget through a direct grassroots participatory process.

This process presents an opportunity for **climactivists** to directly fund efforts to reduce carbon emissions and **bend the climate curve** at the local and state levels.

Permatecture

Design that incorporates biological principles, materials and techniques into man-made structures or landscapes.

See Biotecture

Personal Carbon Allowance or Budget

A proposal developed in the UK to reduce consumption of products containing carbon.

Each person would be given a carbon allowance or ration that would be set at a level that would substantially reduce carbon emissions. The allowance would continue to decrease over time.

See TEQ

Planetary Boundaries

Nine worldwide systems have been identified as having paramount importance to the overall health and integrity of the planet. The boundaries of these systems help define how far we can go before they reach a risk of failure.

These systems are: climate change, biodiversity loss and species extinction, stratospheric ozone depletion, ocean acidification, nitrogen and phosphorus flows, land use changes, freshwater use, atmospheric aerosol loading, and the introduction of novel entities (such as radioactive materials and organic pollutants).

Currently, it is believed that the safe boundaries of climate change, biodiversity loss, and nitrogen and phosphorus flow have been crossed.

http://www.stockholmresilience.org/research/planetary-boundaries/planetary-boundaries/about-the-research/the-nine-planetary-boundaries.html

Planetary Hospice

The planetary hospice movement is intended to provide comfort for the billions likely to experience distress, disruption and death in the coming decades. The goal of the planetary hospice movement is the alleviation of this pain, suffering and panic.

The principles of medical hospice that can be extended to planetary hospice include unconditional support of people, placing priority for pain relief over cure, maximizing person-to-person contact, non-judgmental acceptance, and extensive training for hospice staff.

See Climadelic Therapy

Planetary Personhood

The concept that nature has a legal right to the protection of its integrity.

Those advocating for planetary personhood argue that if non-living entities such as corporations can be considered persons, surely living systems should also be given personhood protection.

Such an idea, if incorporated into the legal doctrine of nations and international bodies, would help protect against both local and international threats (such as climate change) to the integrity of nature.

Ecuador's constitution protects nature, and New Zealand has granted personhood to a forest and river system.

http://www.dailykos.com/story/2014/03/29/1288245/-Planetary-Personhood

https://theconversation.com/what-if-nature-like-corporations-had-the-rights-and-protections-of-a-person-64947?utm_medium=email&utm_campaign=Latest%20from%20The%20Conversation%20for%20December%2023%202016%20-%206383&utm_content=Latest%20from%20The%20Conversation%20for%20December%2023%202016%20-%206383+CID_24203ed9c51ee88b6639225a631ed478&utm_source=campaign_monitor_us&utm_term=What%20if%20nature%20like%20corporations%20had%20the%20rights%20and%20protections%20of%20a%20person

Plenitude

A book authored by Juliet Schor in 2011 describes a small scale post-consumer society with leisure, services, and sharing that minimizes consumption and maximizes fulfillment. *Plenitude* is a useful model for organizing a society in a **post normal climate**.

See Bright, Dark and Light Green Environmentalism, Degrowth

Pollution-Based Prosperity

A pejorative term for one of the pillars of prosperity. It is a false prosperity, as it comes at the expense of the health of living creatures, including us.

See Natural Agricapital, Natural Balance Sheet

Precautionary Principle

Any product, process or substance whose impacts on people or the environment are unknown should be closely examined prior to allowing the product, process or substance to be introduced is the core of the precautionary principle.

A strict application of the precautionary principle to climate chaos would have prohibited the introduction of fossil fuel burning in the nineteenth century until clear evidence of its long-term safety could be established.

Those who would argue the impracticality of that action are confronted with the apparent impracticality of avoiding the decimation of most all life on earth due to the introduction of fossil fuels.

Procrastination Penalty

The penalty imposed on all of us for our collective inaction in drastically reducing **greenhouse gas** emissions is steep, given that most greenhouse gases build up over time and then remain in the atmosphere for many years.

One analysis of the consequences of delaying steep emissions reductions suggests that a delay of ten years requires a subsequent doubling of emissions reductions to provide an equivalent climate impact.

Those who counsel a more deliberative pace of greenhouse gas reductions do not, or choose not, to understand that delay leads to higher concentrations of greenhouse gases and that in turn raises the likelihood of reaching **tipping points** leading to **runaway climate change**.

The term was coined by atmospheric scientist Dr. Michael E. Mann.

https://www.climatecommunication.org/wp-content/uploads/2011/08/presidentialaction.pdf

---

Rebound Effect

When energy savings are partially offset by behavioral changes leading to increased consumption of energy.

Rebound effects can be direct, indirect, system-wide or induced. Direct effects result from consuming more of the commodity whose price has been reduced. A more efficient light bulb results in lower electric costs, which encourages the purchaser to buy more lights or to leave them on for longer at no extra net cost.

An indirect effect would be to take the money saved from having the more efficient light bulb and use it to buy an additional pair of shoes.

A system-wide effect would result from using the cost savings to attend a conference where one gets an idea to start a new business that generates carbon emissions as a byproduct of the business's operations.

And an induced effect would be that energy savings lowers the demand for the fuel. This in turn lowers the price of the fuel, which allows others here or around the world to buy more fuel.

While the existence of the rebound effect is not questioned, these effects are very difficult to quantify or even identify. And thus, the degree to which efficiency gains are offset by rebound effects, while likely to be significant, will vary, and cannot be known with any precision.

See Backfire, Carbon Maze

http://www.newyorker.com/magazine/2010/12/20/the-efficiency-dilemma

---

REDD

A large scale United Nations program, **Reducing Emissions from Deforestation and Forest Degradation** (REDD) has been in operation for more than a decade and contributes to the preservation of **carbon forests.**

Renewables

The holy grail, the promised land, the magic bullet. Renewables are seen as The Answer to climate chaos. And they certainly are a good bit of the answer. Their operation is very low carbon, their energy sources — wind and solar — are inexhaustible and aren't degraded by any known **climate positive feedbacks.**

They tend to capture the public's imagination. Their appearance — sleek solar roof top collectors, futuristic solar panel arrays glistening in the 100-degree desert heat, majestic wind turbines towering over landscapes of corn and soybeans or emerging out of ocean waves — is often dramatic and appealing. They are unlikely to be confused with an oil well, a refinery or a large bellowing power plant.

Yet the reality behind the shiny image is murkier. Renewables currently only deal with a small part of our energy economy. Electricity, which is the easiest to power with renewables, accounts for only 20% of US energy use. Renewables often are best produced in locations — the high plains or deserts — that are remote from most users. And they're less **energy dense** than fossil fuel, requiring much more land and have a lower operating efficiency. There's the tricky issue of **intermittency**, the need for the sun to almost always shine and the wind to almost always blow. And many experts are skeptical that the chemistry and physics of materials will enable us to easily replace the fossil fuels powering ships, airplanes and large trucks.

Even assuming the above issues are addressed, there's the bootstrapping problem. Where do you find the energy to build the massive renewable infrastructure we need without further scorching the planet? Or as energy analyst Richard Heinberg has noted:

> So, in effect they (renewables) are functioning as a parasite on the back of the older energy infrastructure. The question is; can they survive the death of their host?

Lest you think that's a minor issue take a look at these numbers from energy analyst and author Alice J. Friedermann.

> Just one two megawatt wind turbine needs 1,300 tons of concrete, 300 tons of steel, 48 tons of iron, 24 tons of fiberglass, four tons of copper, and so on. And you need about a million of them to provide half our power. And then after 20 years, you'd have to replace them all over again.

Even if we stipulate that we can overcome these and many other non-trivial obstacles (the scarcity of required rare elements, regulatory approvals, growing opposition from legacy utility companies, and financing for a start), do not think it's time to take a climate victory lap.

Carbon emissions come from many sources other than fossil fuels. Deforestation, agriculture, fires, **albedo** decline and methane leaks may not be quickly improved by substituting renewables for fossil fuels. And so, as daunting as it may sound, we cannot simply swap out fossils for renewables and call it a day. We still have to almost entirely transform our systems of food, shelter, transport, production, and more.

For all these reasons, the transition to renewables is a whole lot more complicated than most would expect — but still essential.

See Newables

https://www.peakprosperity.com/podcast/100873/alice-friedemann-when-trucks-stop-running

http://www.rollingstone.com/politics/news/the-koch-brothers-dirty-war-on-solar-power-20160211

Renewable Natural Gas

A gas produced from the decomposition of natural matter with a purity equal to conventional natural gas. Renewable natural gas is being increasingly used to power vehicle fleets, trucks, and buses. Their methane emissions are likely to be lower than conventional natural gas because the production process is both simpler and more controlled.

Resilience

Perhaps no climate related concept is cited more than resilience. And for very good reason. Resilience is the quality that allows for challenges to be experienced without being defeated. We take the blow and then bounce back stronger.

To survive the psychological, economic and physical stresses that climate change will inflict upon all of us, the qualities that comprise resilience are essential at the personal, community and national level.

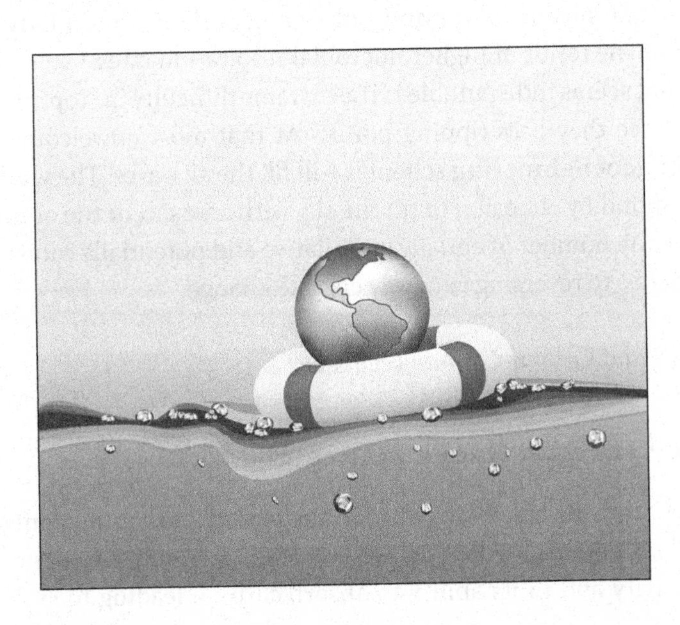

The danger of too great an emphasis on resilience is the probability of moral hazard; that is the temptation to take greater risks as a result of the confidence that one gets from knowing that resilience provides at least a partial insurance policy if things go bad.

There's a vast literature on what resilience is and how to get more of it. Have a look at it.

Runaway Climate Change

When climate chaos occurs exceedingly rapidly as a result of **climate positive feedbacks.**

The most likely cause of runaway climate change would be the interaction between several **climate positive feedback** mechanisms, or **hyperfeedbacks.**

These feedbacks could result from any number of observed or yet-to-be-identified phenomena. These could include the release of methane from permafrost, the release of carbon resulting from the continued decline in the **albedo** in polar areas due to sea ice melts, the loss of tree cover and soil carbon storage capacity in the Amazon and other forests, or any number for other phenomena, such as widespread **bog burns** and forest fires. (A November 2016 study concluded that soil is a very significant *source* and not *sink* of carbon, particularly in the Arctic, as the result of higher microbial respiration rates.)

What seems indisputable is the extreme difficulty in stopping these forces once they pass tipping points. At that most unwelcome point, cries for **geoengineering** schemes will fill the airwaves. The world may well respond by choosing to fill the sky with aerosols or the ocean with iron, or any number of equally speculative and potentially catastrophic approaches to reversing runaway climate change.

See Land Carbon-Climate Feedback

Savory Holistic Range Management

The Zimbabwean biologist Alan Savory argues that the reintroduction of wild animals onto prairies can lead to an improvement in soil productivity and in its ability to absorb carbon, leading to significant drawdown of carbon from the atmosphere (as well as alleviating any climate chaos based **micro guilt** that a meat eater might have).

Savory's approach is both promising and controversial. While rigorous independent evidence for his approach is limited, many of his practices are based on long-standing management practices that ranchers in countries around the world have utilized in a sustainable manner.

Savory's work will be challenging to replicate worldwide in any time frame short enough to provide measurable climate benefits, given the need for substantial local knowledge of soils, weather, topography, animal husbandry, among other factors. Nevertheless, every little bit of carbon reduction helps, particularly if it provides local benefits such as those the Savory approach promises.

The Savory Institute is one of 11 finalists in the Virgin Earth $25 million contest to select the best proposal to remove carbon from the atmosphere.

See Agroecology, Climate Solution Contests, 4/1000 Initiative, Terra Preta

---

Scientific Reticence

Scientists by their training and culture tend to be quite cautious in proclaiming the significance of a finding or in discussing the practical real world consequences of their research. Unless a study shows a greater than 95% confidence that the result is not due to chance it is usually dismissed as not significant.

This can result in a perception of climate chaos as manageable even if the actual studies suggest a much direr outlook. As a result, the public and decision makers may not be getting the full story of the magnitude and likelihood of severe climate impacts.

Using technical jargon, sprinkling the word 'may' in every conclusion, asserting the obligatory 'more study is needed', waiting for months until publication in an obscure journal before sending out a press release may all be standard practices when one is researching the history of religious artifacts in fifth-century Mongolia, but is rather questionable when one has results that could affect the future of 7.2 billion people.

Scientific reticence was first discussed by former NASA climatologist James Hansen in 2007. Hansen has since co-published articles in discussion journals where articles are published and open for peer review at the same time.

A particularly sharp charge supporting the concept of scientific reticence was made by the prominent British climate scientist Kevin Anderson. He has charged that many climate scientists pull their punches when advising governments about the seriousness of the climate situation. Given the grave implications of his charges, it is important to present the entire abstract of the commentary he prepared in 2015:

> The commentary demonstrates the endemic bias prevalent amongst many of those developing emission scenarios to severely underplay the scale of the 2°C mitigation challenge. In several important respects the modelling community is self-censoring its research to conform to the dominant political and economic paradigm. Moreover, there is a widespread reluctance of many within the climate change community to speak out against unsupported assertions that an evolution of 'business as usual' is compatible with the IPCC's 2°C carbon budgets. With specific reference to energy, this analysis concludes that even a slim chance of "keeping below" a 2°C rise, now demands a revolution in how we both consume and produce energy. Such a rapid and deep transition will have profound implications for the framing of contemporary society and is far removed from the rhetoric of green growth that increasingly dominates the climate change agenda.

http://csas.ei.columbia.edu/2016/03/24/dangerous-scientific-reticence/
http://kevinanderson.info/blog/duality-in-climate-science/

Seaweed Farms

Areas on or adjacent to seas where seaweed is cultivated in large quantities.

Studies have shown that seaweed cultivation can absorb vast

quantities of carbon while providing food, medicine, roofing materials and other benefits while helping to combat ocean acidification. It even appears that seaweed fed to cows can reduce or even eliminate the methane content of cow burps and farts.

However, the increasing damage to the world's oceans has degraded seaweed areas and threatens to negate the potential benefits of seaweed farming. The race to expand seaweed farming while the oceans are still healthy enough may be one of the more unsung yet consequential events of our **post normal climate** era.

See Help Kelp! Marine Heat Wave

https://www.inverse.com/article/24093-seaweed-could-reduce-methane-content-in-cow-farts?amp&

Sharing Economy

The movement towards the sharing of goods and services has been facilitated by computerization, GPS, smart phones and hastened by economic weakness. Sharing may have an important role to play in fighting climate change by reducing waste and consumption of resources such as cars, hotel rooms, tools and many other goods and services.

While the hype around sharing focuses on the new, even revolutionary nature of sharing, it is instructive to remember that services such as taxis, libraries, schools, buses and restaurants are all examples of the sharing of goods, services or facilities.

While sharing an extra bedroom rather than booking a hotel room or using a shared car service in lieu of purchasing a car can appear to be quite resource efficient, we still have to grapple with the **rebound** or boomerang effects.

How many more trips to Paris or elsewhere will you take if your shared apartment in Paris is much less expensive than a comparable hotel room? And even if you don't travel more, you'll have extra cash in your pocket that might be spent on any of thousands of goods and services, most of which generate carbon emissions.

Similarly, it is unlikely that the owner of the apartment listed on Airbnb will stuff the extra income under his or her mattress. Rather it will be spent on goods and services that inevitably generate more emissions.

In the absence of a binding carbon cap that limits the overall generation of carbon emissions we can have little confidence that these shared goods and services are fighting climate change, rather than inadvertently accelerating it.

Six Americas

Research performed by Yale University and George Mason University since 2008 has shown that Americans fall into one of six categories — the Six Americas they call them — regarding their attitudes towards climate change.

The classification of the Six Americas in 2015:
Alarmed 17%,
Concerned 28%,
Cautious 27%,
Disengaged 7%,
Doubtful 11%,
Dismissive 10%.

How will the size of each of these categories change over time? Will the doubters and dismissives change their beliefs before their lives are threatened by climate chaos, or even after?

http://climatecommunication.yale.edu/about/projects/global-warmings-six-americas/

Social Cost of Carbon (SCC)

A calculation of the economic and other indirect costs of carbon pollution.

This calculation is used by the Federal government as part of its investment and regulatory decision making. In 2016 the Federal social cost of carbon was $37 per ton. A recent Stanford study however estimated the social cost of carbon to be $220 per ton. This shows the degree to which the magnitude of potential carbon damages is dependent on assumptions and methodology.

This concept of SCC has been criticized from the right and the left. The right sees the social cost of carbon as social engineering whose arbitrary numbers tilt the regulatory process in the direction of more expensive, comprehensive and onerous regulations that penalize initiative and limit freedom.

The humanist or moral critique rebels against the concept that carbon pollution damages can be reduced to dollars and cents.

How can a dollar value be placed on a country that could be destroyed by climate chaos? And what value is placed on the fear and stress that climate change will evoke in a young child?

Many argue that the social cost of carbon is almost infinite, as the consequences of unrestrained carbon pollution is the likely decimation of life on earth.

While these criticisms have some merit, a sufficiently high social cost of carbon will prevent some projects and programs that generate harmful levels of carbon from going ahead as originally designed. The Trump administration cancelled the use of the social cost of carbon in 2017.

https://www.epa.gov/climatechange/social-cost-carbon
climatetest.org

http://www.thedailybeast.com/articles/2016/08/30/the-social-cost-of-carbon-is-the-most-historic-climate-change-decision-yet.html

Social enterprise

A business whose purpose is to bring about positive change while remaining economically viable.

Social enterprises, whether profit or nonprofit, are an increasingly common and viable model for organizing the provision of goods and services in a sustainable and low carbon manner.

See C Corps

Social Proof

Perhaps the non-**climactivist** public's passivity towards climate change is readily understandable. The concept of social proof may explain this passivity.

Social proof focuses on how we are influenced by the behavior of others. As George Marshall says in *Don't even think about it: Why our brains are wired to ignore climate change*:

> Climate change is a global problem that requires a collective response and so is especially prone to this bystander effect. When we become aware of the issue, we scan the people around us for social cues to guide our own response: looking for evidence of what they do, what they say, and, conversely, what they do not do and do not say.

And so, if the people around us aren't acting — whether to stop a mugging or to fight climate change with vigor — we tend not to either.

If climate chaos is so potentially devastating and even an existential threat to all of life, why is it that most of our leaders in politics and the media, in business and in academia treat it as an important issue but not one that rises to the urgent — much less **eemergency** — level? If leaders aren't acting like it's an eemergency then most of us will also act like it is not an **eemergency**. Simple, isn't it?

Take former President Obama. He initiated dozens of policies addressing climate chaos. He gave speeches, attended the Paris COP21, negotiated a climate deal with China and more. But did he address the nation from the Oval on climate chaos? Did he call a leadership summit of legislative leaders or call for bold new climate legislation since early in his administration? Did he proclaim that there is a true climate emergency?

No, no and no. So even the most loyal Obama supporters would have had no reason to believe that climate chaos requires urgent much less eemergency attention. And that is social proof.

Socialized Soot

A phrase used by former Congressman Bob Inglis as part of his argument for action against climate chaos based upon conservative values and policy choices.

Socialized soot refers to those companies or individuals that spread soot and other pollutants through the atmosphere as a byproduct of production and do not have to the pay the costs that the rest of us who breathe in the soot incur. The benefits of the socialized atmosphere accrue to the polluter and the larger public pays the aesthetic, health and environmental costs.

http://midwestenergynews.com/2014/06/27/qa-the-free-enterprise-case-for-climate-action/

Solar Radiation Management (SRM)

A form of geoengineering that attempts to reflect incoming solar radiation back out to space. Two common approaches to SRM involve spraying seawater to brighten clouds or injecting aerosols into the upper atmosphere.

See Geoengineering

Solastalgia

Profound or existential distress caused by environmental chaos, particularly close to home. The word was created by Glenn Albrecht in 2003 from the words for solace and nostalgia.

See Ecocholia

Sources and Sinks

This term refers to the many pathways that distribute CO_2 and other greenhouse gases into and out of the atmosphere. Sources send CO_2 into the atmosphere, sinks absorb it from the atmosphere.

Climate chaos can be looked at as an imbalance between sources and sinks. Fossil fuels and cement manufacturing account for about 75% of the sources of CO_2 with deforestation and other land uses accounting for much of the remaining 25%. Sinks include photosynthesis by plants on land and phytoplankton and other chemical reactions occurring in the seas.

See Carbon Clogs, Land Carbon-Climate Feedback

Steroid story

The influence of climate chaos on any given weather event is endlessly debated. One useful analogy is to compare a weather event to a baseball player who uses steroids. The steroids make the player stronger and more likely to hit a home run. But any one home run he hits cannot be definitively attributed to steroids, as he hit home runs before he took steroids but many more since his use of steroids.

So, too, any severe weather event is more likely to occur due to the steroid-like increase in climate activity due to global heating, but any one event may or may not have occurred or been at the same strength.

And that's the steroid story — at least for now. As our scientific knowledge increases and as events become more severe and more frequently outside historical norms, climate scientists are beginning to be able to attribute with more confidence a given weather event to climate chaos.

Stocks and Flows of CO_2

An important distinction in atmospheric science is between the stock or the total amount of CO_2 in the atmosphere and the flow or amount added to or taken away from the stock over a period of time.

CO_2 remains in the atmosphere for decades. Reducing the flows of CO_2 into the atmosphere as the world is attempting to do will thus have no immediate effect as a result of the large and stable stocks of atmospheric CO_2 that will remain for many more years.

See Carbon Lock-in, Legacy Emissions, 2° Dogma

Subprime Carbon

The efforts to **Keep It in the Ground** and prevent carbon from being burned has led some including Al Gore to label carbon as sub -prime. Subprime as its value may plummet if carbon in the form of fossil fuels can be successfully kept off the market. And subprime because it's a toxic and risky material that endangers life on earth.

Sunny Day Floods

High water tables can lead to flooding, even when there is no active precipitation. Miami Beach is one such place with sunny day floods given its high-water table and porous limestone soils. Since flooding is occurring with less than a foot of sea level rise, imagine — if you dare — what the streets will look like with a sea level rise of up to 6 feet, as has been predicted by some studies. The title of this entry might then simply be **Sunny Day Without Flooding**, as sunny skies without flooding would be rare enough to rate an entry.

See Global Weirding

Technohubris

The belief that technology can overcome virtually any kind of human or natural problem or issue in modern society. Our technohubris appears to be destroying a habitable planet by acting as if fossil fuels and endless GDP growth are benign no matter how destructive their long-term effects may be.

Termination Shock

An abrupt end to certain **geoengineering** activities can result in an even worse situation or termination shock than before the geoengineering began. Termination shock is similar in principle to the risk of abruptly stopping certain medicines. The termination danger can result from the fact that greenhouse gas continues to build up in the atmosphere in many geoengineering schemes, so that, if the scheme is ended, higher levels of greenhouse gases are unleashed.

As a result, most proposed geoengineering needs to be in place indefinitely, something hard to guarantee given expected political, economic and climate instability.

Terra Preta

Extremely humus rich highly fertile dark soil that can store large amounts of carbon.

This soil, found in the Amazon as well in Liberia and Ghana, is created by mixing biochar (a kind of charcoal) with organic material. The use of terra preta can significantly increase carbon absorption and may modulate global heating. This technique, which may be have used as long ago as 5,000 years, is likely to become a key part of the carbon farming and **agroecology** movements.

See Bio Sequestration, 4/1000 Initiative

Third Pole

The mountains and high plains in East Asia are often called the Third Pole as a result of the massive glaciers they contain. This area includes the Himalayas, the Tibetan Plateau and the Hindu Kush ranges.

Some 1.3 billion people, nearly one in five of the world's population, depend upon the 10 river systems that originate in the Third Pole. An analysis of satellite photos in 2016 shows that about 10 inches of water has been lost annually throughout the Himalayas since 1973, a staggering loss of water.

As these glaciers continue melting, and ice famines get progressively worse, water supplies for hundreds of millions of people are being seriously jeopardized.

http://www.bbc.com/news/science-environment-38307176?utm_source=Daily+Carbon+Briefing&utm_campaign=cd7269fdad-cb_daily&utm_medium=email&utm_term=0_876aab4fd7-cd7269fdad-303479513

Tipping Point

The situation where the environment is on the verge of changing to a different state. A small stimulus triggers a larger change and the state of the system is transformed. A potential tipping point can be catastrophic and difficult to accurately anticipate in advance.

Potential tipping points can result from methane gas release from permafrost and the seas, ice loss from the Arctic or the West Antarctic Ice Sheet, deforestation of the Amazon, and many others sources.

Multiple tipping points can lead to highly unpredictable and unstable **hyperfeedbacks**.

A November 2016 study examined a series of 37 potential climate tipping points and reached a very sobering conclusion:

> One of the most important findings is that 18 out of 37 abrupt changes are likely to occur when global temperature rises are 2° or less, often presented as an upper level of "safe" global warming. Our results imply that there is no window of "safe" global warming and no threshold separating safe and dangerous climate change. Every 0.5° temperature increase is similarly dangerous. Little or nothing can be done to reverse a tipping point once it is underway.

Social and economic tipping points may perhaps be as important as physical tipping points in addressing climate chaos. The time when enough people in a community or nation accept the need for urgent climate action will be a critical tipping point. As wind and solar become competitive with fossil fuels a decisive tipping point in favor of these renewables may not be far away.

See Climate Panic Points, Climate Positive Feedback, Land Carbon-Climate Feedback, 2° Dogma

https://theconversation.com/what-climate-tipping-points-are-and-how-they-could-suddenly-change-our-planet-49405

Topocasm

Theodore Gaster's word for the entire locality — soils, plants, animals, everything, yesterday, today, and tomorrow — taken as an organism. The topocasm will change in a **post normal climate** with unknown consequences.

Tradable Energy Quotas (TEQ)

A rationing proposal developed in the United Kingdom a decade ago.

The TEQ is similar to a carbon **cap and trade** system applied both at the individual and institutional level. Each person is given a weekly quota in electronic form that can be redeemed for purchases based on the carbon content of each purchase. Over time, the overall national and personal quotas decline to reflect a progressively shrinking carbon budget.

The proposal won praise in the UK environmental community but to date has not received any significant political support.

http://www.teqs.net

Transport Triage

The transport sector is surprisingly wasteful. Think of cars. Most are used for perhaps an hour a day and sit idle for the rest the time. With the world soon to have one billion cars, we waste 23 billion hours of potential use every day or an average of three hours a day for every person on earth. This solitary confinement also requires a space for storage — a garage, part of a street or a lot. Imagine the financial and climate related cost of billions of mostly unused asphalt or concrete paved parking spaces.

A 160-pound person in a car that weighs 4000 pounds means that there is 25 times as much vehicle mass as the mass of the person being transported. So even occasional carpooling, some degree of car sharing and alternative fuel powered cars cannot make the automobile a particularly efficient means of transport in a **post normal climate** era.

Some form of triage to drastically transform and downsize the world fleet should occur.

See Billion Car Challenge

Under2 Coalition

The Under2 coalition, founded in 2015 by California and Baden-Württemberg, Germany, brings together over 165 cities, states and countries committed to either reducing greenhouse gas emissions equivalent to 80 to 95 percent below 1990 levels or to less than 2 metric tons per capita by 2050. These jurisdictions cover six continents and represent over a third of the global economy.

Currently, 10 US states and eight cities are members.

Vertical (Ocean) Farming

A concept developed by long time fisherman Bren Smith layers the cultivation of fish and sea animals through the use of a carefully designed and placed sea installation generally extending from 20 to 500 feet deep.

The advantage of vertical ocean farming is that a wide range of fish can be cultivated given the variety of niches. Vertical farming may be called upon to provide larger fish catches as traditional ocean productivity continues to decline. However, scientists fear that as areas of the ocean become too toxic to practice aquaculture vertical farming may be adversely affected.

Various land-based vertical farm proposals are also being considered, from vertical farm skyscrapers to small scale vertical food growing systems.

Vicious Cycle See Climate Positive Feedback

Virtual Power Plant (VPP)

A network of decentralized power sources that functions as a unified system.

Instead of building large power plants, whether powered by renewals or fossil fuels, software is beginning to link decentralized sources of power such as rooftop solar, microgrids, and car and home batteries. Linking VPPs with sensors and software to maximize energy efficiency will allow these systems to power our homes, cars and factories with greater efficiency, less concern for **intermittency** and less use of fossil fuel.

Warm Arctic, Cold Continents

A name some scientists have given to the not yet fully accepted understanding that the considerable warming of the Arctic results in more cold weather lower in latitude due to changes in the jet stream and in other atmospheric flows. One possible consequence of this phenomenon, should it continue, will be to give ammunition to climate deniers who use the specious argument that cold winters show that climate change is not real.

See Arctic Amplification

https://www.washingtonpost.com/news/energy-environment/wp/2016/12/23/the-arctic-is-behaving-so-bizarrely-and-these-scientists-think-they-know-why/?postshare=5831482554415631&tid=ss_tw-bottom&utm_term=.b0e12edad7b2

Watermelon Snow

A species of green algae with pink color that thrives in icy water and is common in polar and alpine regions. Its color gives it a lower **albedo**, leading to greater heat absorption. This melts more snow leading to the likelihood of more algae, more watermelon snow and ultimately greater global heating.

See Climate Weirding, Positive Climate Feedbacks

Whole Cost Accounting

Accounting will play a central role in the climate chaos drama. How society decides to count and value carbon and related pollution will influence what kind of investments and other expenditures corporations, small businesses and individuals make.

> While there may be no 'right' way to value a forest or a river, there is a wrong way, which is to give it no value at all. How do we decide the value of a 700-year-old tree? We need only to ask how much it would cost to make a new one, or a new river, or even a new atmosphere.
>
> Author Paul Hawken

Whole cost accounting attempts to value the full costs of an asset, liability or a transaction whether or not there is a clear financial value.

Zombie Arguments

The myths circulated by climate deniers that keep recycling themselves despite repeated debunking. Among these myths are that climate scientists themselves pronounced global warming as a hoax, or that the infamous hockey stick graph showing the increase in atmospheric temperatures over centuries is a fraud.

https://www.poynter.org/2016/despite-fact-checking-zombie-myths-about-climate-change-persist/443460/

Zone Creep

The geographical zones on earth — tropic (torrid) zones, temperate zones and frigid (polar) zones areas — are all shifting.

The polar zone is shrinking while the temperate zones are migrating northward in the northern hemisphere and in the southern direction in the southern hemisphere. This is one of the many instances of the modification of the **climate chronosphere**.

See Biotic Refugees, Reef Races

New Climate Terms

These terms are not currently in use, or are used only in a non-climate change context.

Active Infrastructure

Infrastructure that supports physical activity and mobility while minimizing the generation of CO_2 emissions.

Active infrastructure may include paths, sidewalks, pedestrian streets, bikeways and traffic separation structures. Parks, playgrounds and gardens can also be thought of as active infrastructure. Context, design, location, accessibility, safety and many other qualities will determine whether this infrastructure actually is effective in enhancing low carbon mobility.

See Automated Movement Networks, Billion Car Campaign, Motion Generated Energy, National Climate Defense Interstate Mobility Transportation System, Transport Triage, Universal Climate Design

---

Ameliorist

Someone who believes that climate chaos can be successfully fought without major societal mobilization and with little effect on the economy or society. The economist and columnist Paul Krugman was a prominent ameliorist. A 2014 column of his concluded as follows: "If we ever get past the special interests and ideology that have blocked action to save the planet, we'll find that it's cheaper and easier than almost anyone imagines."

Recent comments by Krugman reflect a somewhat more urgent tone.

See Brightsiders, Bright, Dark and Light Environmentalism

http://www.nytimes.com/2014/09/19/opinion/paul-krugman-could-fighting-global-warming-be-cheap-and-free.html?_r=0

Americans for Climate Prosperity

A proposed nonprofit advocacy organization with the mission of enhancing prosperity, while fighting climate change. Its name stands in contrast to Americans for Prosperity, a Charles and David H. Koch sponsored climate denying organization with vast influence within the United States.

See Degrowth, Plenitude

Anticipatory Migration See First Flee/er

Anticipatory Reparations

A proposal to provide financial and technical **readaptation** support to those underdeveloped countries anticipated to experience the most climate chaos related damage. Providing education, infrastructure, and other support well before damages occur can prevent or minimize these damages.

The word *reparations* (rather than payments) is chosen as developed countries largely created the climate crisis through decades of carbon emissions in the service of economic development (and some would say war and exploitation of underdeveloped countries). Developed countries have rejected climate aid commitments labeled as or considered to be reparations, anticipatory or otherwise.

See Climate Debt

Appliance Euthanasia

Eliminating appliances while they are still working to save energy, upkeep cost or space.

As climate chaos intensifies, emissions reductions are mandated and incomes fall, euthanasia will inevitably be performed on other consumer goods and services as well.

To hasten and accept appliance euthanasia, we may need to create a National Find Other Means for Gratification Project, perhaps based on **plenitude** or **degrowth**.

Arctic Ice Sensitivity or Claim Your Arctic Ice While It Lasts

A remarkable albeit sobering study has determined that every ton of CO_2 burned anywhere in the world results in the melting of 32 feet of Arctic summer ice. That means that the average American burning 16 to 20 tons of CO_2 every year melts summer sea ice equal to the area of a one bedroom apartment. So, the next time you get on a plane or swap your 40-inch flat screen for a 55-inch model think of those polar bears being stranded on an ever- shrinking pad of ice. And don't forget the greater climate chaos resulting from the loss of ice due to **climate positive feedbacks** and **albedo** loss.

See Albedo, Guilt Per Gallon, Hyperfeedback, Micro Guilt, Polar Bear Propaganda

http://www.theatlantic.com/science/archive/2016/11/the-average-american-melts-645-square-feet-of-arctic-ice-every-year/506441/

Association of Climate Mensches

Those individuals with decency, integrity and essential niceness who address the challenges of climate chaos.

If we have climate heroes, villains, deniers, doomers, ameliorists, realists, neoskeptics, gradualists, brightsiders, skeptics and climactivists don't you think there ought to be recognition for mensches as well?

Automated Movement Networks

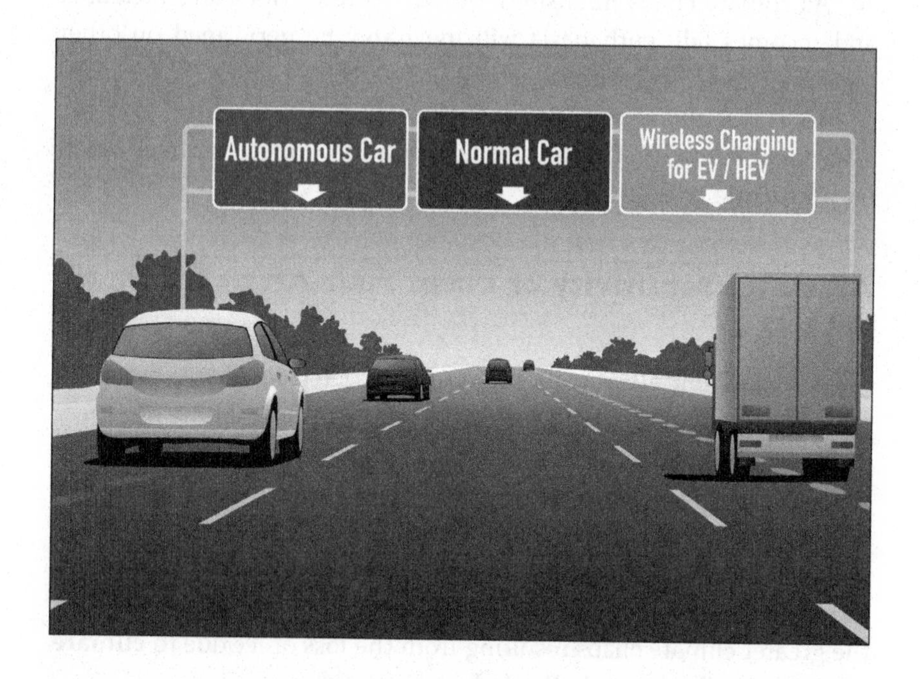

A suite of technologies is emerging to revolutionize how we travel, particularly in urban areas. This emerging automated network combines GPS, self-driving, electric propulsion, and car sharing.

While the impacts of these advances are difficult to predict with certainty, it is likely that car ownership will decline, traffic will run smoother, most parking areas will be converted to other uses and accidents will become increasingly rare. Most importantly, vehicle miles travelled should decline and carbon emissions will be reduced.

See Billion Car Campaign, National Interstate Climate Mobility System, Roads Closed Ahead, Transport Triage

Baseline Bias See Placebo Plan

Behavioral Tipping Points See Planetary Panic Points

Bifocal Behavior

The capacity to focus on short and long term concerns simultaneously.

Most would agree that our collective behavior is skewed to the short term. There's a seemingly unshakable devotion to a corporate time line defined by quarterly performance and a political time line defined by elected officials calibrating actions for their impact on the next election. And a noticeable absence of urgency to act on climate chaos.

Can anyone think of an example of our society having too much long-term thinking? When will political leaders focus on the period beyond the next election? And when will the corporate quarterly report refer to a quarter of a century instead of a quarter of a year? The answers to these questions may determine our ability to preserve a habitable planet.

See Distraction Dilemma, Pagency, Patient Climate Capital

Big Wisdom

A process of aggregating and integrating the wisdom that large numbers of people possess in order to gain insights that are beneficial to society.

Big wisdom is distinct from big data, which is a process of aggregating massive amounts of data to gain insights that would otherwise be difficult to ascertain. We need both big data and big wisdom as well as the discernment to integrate the two.

Biocodes

Biologically based building codes.

Current building codes may not fully encourage design practices and features based on natural systems and living biology.

These features and processes include green walls and roofs, passive solar design, geothermal heating and cooling, natural ventilation, natural and full spectrum light, smog eating materials and many other cutting edge natural technologies. Biocodes promise to enhance environmental and climate benefits that current nonliving materials and processes may not be able to provide.

Bio emissions vs Techno emissions (Cows vs Cars)

Bio emissions result from digestive activity by animals, such as methane releases from cows or cow waste (cow farts or manure). California is among the first jurisdictions to attempt to regulate bio emissions through capturing methane released by cows and preparing special feeds, possibly including seaweed, to reduce cow gas. One feed, made of biochar, citric acid and garlic claims to reduce cow methane emissions by 30%. According to the company that produces this product, feeding every cow their feed would reduce CO_2e emissions equivalent to taking 200 million cars of the road.

It's not clear whether cows will be more pleasant to be around as the reduction in cow farts may be offset by their newly-gained garlic breath.

Techno emissions are those caused by generators, motors and other means of burning fossil fuels.

Controlling both categories of emissions is a prerequisite to maintain a habitable planet.

https://www.inverse.com/article/24093-seaweed-could-reduce-methane-content-in-cow-farts?amp&
http://www.mootral.com

Biofacture

The manufacture of products using biological processes.

This approach could provide products with a much lower input of natural and fossil fuel resources. One example is a current effort to grow bacteria and algae in ways that allow them to be used as the basis for textile and other materials.

http://www.goodnet.org/articles/future-fashion-growing-your-own-clothes

---

Biogrief See Solastalgia

Biomortis

The death of living creatures.
An increasingly common consequence of climate chaos.

See Carbicide

Biophonic Discord

Biophony as conceived by soundscaper Bernie Krause is "all the living organisms that vocalize in a given habitat, sounding together".

Krause found that natural sounds made by living creatures are being disrupted by the impacts of a variety of environmental insults, including climate chaos.

He further argues that the ability of species to thrive depends upon having intact biophonic niches.

Listen to Bernie Krause's TED talk 'The Voice of the Natural World' below at the 11:00 minute mark near the end and pay close attention to the sound of a beaver mourning the loss of its family. Multiply that sadness, that grief by the billions of creatures likely to be harmed by climate chaos and you will have some sense of the magnitude of the loss we are imposing on the plants and animals with whom we share this planet.

See Collective Empathy, Geophony

https://www.ted.com/talks/bernie_krause_the_voice_of_the_natural_world?language=e

Biotecture

The design of buildings and landscapes on the basis of biological processes.

Biotects are professionals who design and oversee the construction of biotectural structures and landscapes.

Biotic Refugees

Virtually every plant and animal species is relocating to attempt to survive climate chaos. Scientists are documenting these species migrations in the hopes of learning how each species adapts to climate chaos. A recent study estimated that land species are migrating over three miles a decade towards colder climates while marine life is migrating an astounding 45 miles a decade. Unfortunately, climate change is outpacing the ability of many, perhaps most, species to adapt.

These migrations are beginning to create economic hardship. Along the Atlantic seaboard, for example, the migration of many fish species to cooler northern waters is disrupting traditional fishing patterns.

Not all species should be assisted in migrating. Perhaps pests and pathogens (such as mosquitos) should be encouraged to stay home, however difficult that may be to achieve.

We can improve the success of biotic refugee migrations through creating wildlife corridors, minimizing land disturbances and restoring ecosystems.

> One of the defining features of the Anthropocene is that the world is changing in ways that compel species to move, and another is that it's changing in ways that create barriers — roads, clear-cuts, cities — that prevent them from doing so.
>
> Elizabeth Kolbert, The Sixth Extinction: An Unnatural History

Of course, humans are biotic refugees as well. And our climate migrations too have begun.

http://www.independent.co.uk/environment/climate-change-global-warming-changing-nature-human-assisted-evolution-a7410286.html
http://science.sciencemag.org/content/354/6313/aaf7671
http://mobile.nytimes.com/2016/12/30/science/fish-climate-change-northeast.html

Biotic SAD (Seasonal Affective Disorder)

Almost every living organism is affected by changes in the duration and timing of seasonal light and temperature (known as phenology). Given the adaptation of species to their environments over eons, as climate chaos affects seasonal characteristics almost all species will be forced to adapt, often sub-optimally or in many cases unsuccessfully, leading to a kind of biotic SAD. The SAD will often arise because of a mismatch in the timing and extent of adaptation by different species.

See Climate Chronosphere

Bird Brain Brilliance

Zebra finches that laid eggs when temperatures were higher than 78° sang a different tune to their unborn chicks than were sung to other eggs when temperatures were lower. As a result, the chicks that were born after experiencing these special songs were smaller and thus were better able to cool down, enabling them to adapt to higher temperatures.

It appears that finches are adapting to climate change better and faster than we are. So, who are the real 'bird brains'?

See Climate Shapeshifting

http://www.theverge.com/2016/8/18/12490292/zebra-finch-climate-change-call-heat-warning

Body Protection Factor (BPF)

A proposal for the creation of a tool to measure the effectiveness of products or processes intended to shield the body from the effects of extreme climate conditions. Analogous to the Sun Protection Factor (SPF) used in selecting sunscreen.

Bog Burns

Peat bog fires that release enormous quantities of carbon into the atmosphere.

These fires are difficult to extinguish and may smolder for months or longer. The recent widespread Indonesian fires affecting 43 million people were estimated to release as much carbon as the entire country of Germany emits in a year. The increasing severity of these bog burns is thought to be a result of higher temperatures and increased dryness.

Bogs cover 3% of the world's land area, and are estimated to contain more than twice as much carbon as occur in all the forests and grasslands, and nearly as much carbon as in the atmosphere. A newly discovered bog in Congo is estimated to contain carbon equivalent to 20 years of carbon emissions generated by the United States. A global initiative to protect peat bogs was launched in 2016.

See Climate Positive Feedback, Disaster Emissions, Man Made Natural Disasters

http://www.wri.org/blog/2015/10/indonesia's-fire-outbreaks-producing-more-daily-emissions-entire-us-economy
http://mobile.reuters.com/article/idUSKBN13C2GM

Breaking the Climate Curve

Ultimately, **bending the climate curve** may not be sufficient to preserve a livable planet.

Breaking the climate curve so that temperatures do not further increase will be required. That means the necessity of taking carbon out of the atmosphere in almost unimaginably large quantities over decades with the goal of restoring carbon concentrations to 350 parts per million or less.

See BECCS, CCS, Carbon Forests, 4/1000 Initiative, Negative Emissions, Terra Preta

---

Buy Once Economy

An economy characterized by products that are designed to be durable and appealing enough to be purchased and kept indefinitely. Reducing consumption is an important way to reduce carbon emissions. One way is through a Buy Once economy. (Some have suggested that the entry should be called the Bey/once economy.)

Carbicide

Deaths attributable to climate chaos.

A 2012 study by the Climate Vulnerability Forum estimated that 400,000 people die from the effects of climate chaos each year, or over 1,000 a day. An additional 4.5 million died in 2012 from what the Forum calls the carbon economy. These numbers are broad estimates and are disputed by some. No doubt these death rates have increased since then, given the increases in temperature, wildfires, migration and overall planetary disruption.

Think about these numbers for a moment; climate deaths are the equivalent of perhaps 4 large jetliners crashing every single day. Every home page and TV network would be screaming about the horrors of flying and the world-wide crisis we are facing. Congress would be holding hearings and politicians would be asked what they are doing to make us safe again. Hysteria would rule.

Yet how many even know that the equivalent of four jetliners crashing every day are dying from climate change?

See Media Omertà

https://www.newsecuritybeat.org/2012/11/climate-change-kill-million-people-year-daras-2012-climate-vulnerability-monitor/

Carbon Bariatrics

The theory and practice of reducing carbon emissions.

Carbon Clogs

Carbon **sinks** are the mechanism for absorbing carbon from the air, and are helping to regulate the overall level of carbon in the atmosphere. Ocean, soil and forests are the main natural sinks. Higher temperatures, drought or excessive rain can all potentially neutralize or stop carbon from being absorbed. This results in the sink becoming carbon clogged, or to put it another way the sink becomes a carbon source.

An example of a major carbon clog can be found in the Amazon, where forests are highly sensitive to rainfall. A recent study demonstrated that droughts in 2005 and particularly in 2010 completely shut down the Amazon Basin's carbon sink through slowed tree growth and tree death.

And although the carbon sink recovered after 2005 and 2010, the next Amazon carbon clog is almost inevitable. And since global temperatures are considerably higher than in 2010 can we count on the carbon sink unclogging next time? No one knows.

The larger concern is that we will discover that carbon clogs become all-too-common a natural response to climate disturbances. This may close off the opportunity to use the capacity of carbon sinks to draw down greenhouse gases from the atmosphere. In addition to damaging the health of these remarkable forest ecosystems, the consequence of clogs becoming **sources** in the Amazon and elsewhere would be the hastening of a **climate positive feedback** that could lead to **runaway climate chaos**.

See Carbon Forests, Carbon Maze, Land Carbon-Climate Feedback, Sources and Sinks

http://www.bbc.com/news/science-environment-36856428

Carbon Complicity Index

Many individuals and institutions (including our national government leaders) are intentionally misleading society about the reality and consequences of climate change.

Creating a carbon complicity index would allow the measurement of the degree of one's complicity in actively misleading society. The index could be used to support the prosecution of individuals for **carbon crimes against humanity.** The index could also be used to bring these folks in front of the proposed **Climate Truth and Reconciliation Commission.**

Carbon Consumption Conspiracy

10% of the population is estimated to generate perhaps half of all carbon emissions.

This imbalance suggests that a strategy for reducing emissions should focus not on the overall population, but on those relatively small numbers of people in the Western world and in the affluent areas of other countries who generate the highest amounts of carbon emissions. Of course, 10% of 7.2 billion people is still 720 million people, which is about equal to the collective population of the US, Russia, Canada, Australia and Brazil.

This segment of the world's population benefits the most from a high carbon lifestyle while disproportionately burdening the rest of the world with its carbon emissions.

One rarely sees this statistic in any discussion of climate action. Perhaps this has something to do with the carbon consumption of people working in the media, government, and corporate sectors. A conspiracy? Perhaps not. But is it reasonable to assume that it is just a coincidence?

See also Climate Debt, Frequent Flyer Fee

https://www.oxfam.org/en/pressroom/pressreleases/2015-12-02/worlds-richest-10-produce-half-carbon-emissions-while-poorest-35

Carbon Cost Shifting

The combustion of carbon imposes great costs on many. Poor countries suffer disproportionately from carbon combustion, as climate chaos has a much greater impact in these countries given their lack of resources to undertake effective adaptation efforts. These countries are also disadvantaged by not benefiting from the affluence created by the burning of carbon now and in past decades.

Poorer communities in more affluent countries also suffer disproportionately due their geographic location and often limited political power.

Future generations and younger populations will receive few of the benefits of carbon combustion while bearing the full and likely catastrophic cost of climate chaos. The columnist Dave Roberts calls this shift in cost to future generations and poorer communities and nations a "geographic discount rate."

One could say: "Never was so much harm done to so many for so long by so few."

See Climate Debt, Contraction and Convergence, National Discount Rate

Carbon Coupling/Decoupling

This refers to the relationship between the increase in emissions relative to the increase or decrease in economic activity, as measured by gross domestic product (GDP).

Until 2014, the increase in world GDP was always accompanied by an increase in carbon emissions. In the past three years GDP has modestly increased while carbon emissions have remained roughly flat, largely due to a reduction in coal-use by China and an increase in renewables throughout the world.

Though some have hailed the apparent end of annual CO_2 emission increases, the situation is like a baseball team that falls further behind each inning. Finally, they go three innings without falling further behind. Yet it's now the ninth inning and they only have one chance left to win the game. And so too the world has only one chance (if that) of winning the battle against climate change and it's right now.

To reduce carbon emissions by 80% or more within the next decade or so as required to remain within the **carbon budget** without a total collapse in the world economy requires a radical decoupling of carbon from GDP.

The rate of necessary carbon emission reductions ranges up to 10% or more a year over a multi-year period, an unprecedented reduction that will be exceedingly difficult to achieve without a severe long term depression or a dramatic breakthrough in national climate commitments. The burden of proof must be on those who argue that decoupling of this magnitude is possible, rather than on those who argue that GDP must be reduced.

See Carbon Budget, Decarbonization Dilemma, Degrowth

http://www.nature.com/nclimate/journal/vaop/ncurrent/full/nclimate2892.html

Climate Hand-Waving

Appearing to make substantive arguments while actually making insubstantial and misleading arguments in an attempt to obfuscate. Based on the term "hand-waving".

So, keep your carbon hand-waving detector handy when evaluating proposed climate remedies, particularly given the varying perspectives that **ameliorists, brightsiders, climate gradualists, eemergency** activists, **bright, dark and light green environmentalists, climate realists, climate doomers** and outright deniers have on climate threats and remedies.

Carbon Junk Food

Higher levels of CO_2 in the atmosphere appear to lead to lower levels of iron, zinc and protein in corn, wheat, rice and soybeans. These nutritional deficits transform healthy foods into junk — or at least much less nutritious — food. Lower levels of protein and higher levels of carbohydrate may also lead to greater obesity.

The mechanism by which this occurs is not known. While it may be a result of plants growing larger because of more CO_2 in the air, recent studies suggest that plants may not always grow faster or larger when CO_2 concentrations are higher.

See Adaptation/Mitigation, Plant Purgatory

http://news.nationalgeographic.com/news/2014/05/140507-crops-nutrition-climate-change-carbon-dioxide-science/

Carbon Maze

The increasingly difficult and obstacle laden path to significant carbon reductions is analogous to navigating a maze or labyrinth, as many of the accepted approaches to reduce carbon concentrations may be on the verge of being foreclosed. Examples follow:

Seaweed cultivation provides food and habitat for sea life and people while absorbing large quantities of carbon. Yet increased ocean temperatures appear to be killing seaweed through bacterial attacks and a reduction in seaweed's immune resistance.

Tree planting is touted as an effective means to absorb CO_2. Yet the increasing frequency and intensity of forest fires as well as chemical changes to the soil as a result of higher temperatures are negating the carbon benefit of forest creation. Intense wildfires appear to make forest regrowth more difficult.

Hydropower is almost universally believed to be a renewable energy source. Yet a recent study by Washington State University finds high levels of methane given off by the storage reservoirs associated with dams; these emissions are not even included in greenhouse gas inventories. Perhaps 'methane maze' needs to be an entry in the *CVF* as well.

Recently the 2016 Kigali Agreement to phase out hydrofluorocarbons, or HFCs was negotiated. Yet HFC's were only introduced into the atmosphere as a result of the 1987 Montreal Protocol that phased out chlorofluorocarbons, or CFCs. (HFC phase-out is projected to bend the climate curve downward.)

Obstacle laden pathways in the carbon maze are created by us as well. One is the paradox that while a great recession or worse may be the only way to significantly reduce emissions we are then likely to see the election of reactionary governments hostile or at least indifferent to the need for vigorous climate action.

Let us hope that sustainable pathways through the carbon (and methane) maze to reach the light at the end of the tunnel (to mix metaphors) can soon be identified.

See Carbon Forests, Carbon Resistance, Climate Positive Feedbacks, Help Kelp, Seaweed Farms

https://insideclimatenews.org/news/21122016/california-forests-wildfires-climate-change
https://www.theguardian.com/sustainable-business/2016/nov/06/hydropower-hydroelectricity-methane-clean-climate-change-study

---

Carbon Nudges

The practice of nudging or subtle behavioral modification was described by Professor Cass Sunstein in his 2009 book *Nudge*.

A successful example of using a nudge to address climate change was demonstrated in a controlled experiment undertaken by Virgin Atlantic Airlines in 2016. The airline provided information on fuel consumption and incentives to its pilots with the goal of reducing fuel consumption.

The experiment was a success with both the pilots receiving information and/or incentives and the control group that did not receive information or benefits saving significant amounts of fuel (the pilots who received incentives and information saved more fuel than the control group thus demonstrating the value of the carbon nudge).

While the world needs much more than a carbon nudge or two, it's easy to imagine a slew of carbon nudges being tried by governments and companies throughout the world. Those spearheading the nudge movement perhaps should be known as carbon nudgeatarians.

http://blogs.worldbank.org/publicsphere/voices/how-virgin-atlantic-used-behavior-change-communication-nudge-pilots-use-less-fuel-reduce-emissions

---

Carbon Resistance

Scientists are observing many instances where compensatory mechanisms are negating efforts to reduce CO_2 concentrations.

Planting trees appears to lead to a reduction in carbon absorption by the world's oceans. Reservoirs created by hydropower dams are found to give off large quantities of methane. And at least one study suggests that painting roofs white to increase the urban **albedo** may not be effective as reflections may lead to greater atmospheric heating.

These and other examples point to a kind of resistance by carbon against efforts to restrain its global heating effects.

See Carbon Maze, Climate Positive Feedbacks

https://web.stanford.edu/group/efmh/jacobson/Articles/Others/HeatIsland+WhiteRfs0911.pdf

https://www.theguardian.com/sustainable-business/2016/nov/06/hydropower-hydroelectricity-methane-clean-climate-change-study

---

Carbon Overshoot

The scenarios for restraining planetary temperature increases to 3.6° rely on overshooting atmospheric carbon concentrations and temperature increases and then initiating massive **carbon drawdown** later in the century. **Negative emission** technologies would then slowly bring temperatures back down over time. At least that's the theory.

See BECCS, Carbon Budget, Carbon Lock-in, Carbon Path Dependence, Carbon Time Travel

Carbon Path Dependence See Carbon Lock-in

Carbon Piracy

As CO_2 emissions become increasingly restricted, the illegal recovery and burning of fossil fuels is likely to increase. While it may be difficult to recover carbon from underground sources without detection, other CO_2 emissions are much more difficult to police, such as slash and burn farming and cattle production, industrial agriculture and deforestation.

See Carbon Dye, Carbon Exclusion Zones, Carbon Lock Box, Climate Justice System

Carbon Police

A specially trained force focused on enforcing laws passed to regulate the use of carbon.

Carbon police will become increasingly common, modeled perhaps on the agency best known for fighting carbon piracy — the Brazilian Institute of Environment and Renewable Natural Resources [Instituto Brasileiro do Meio Ambiente e dos Recursos Naturais Renováveis] (IBAMA).

Carbon Recriminations

As climate chaos becomes increasingly calamitous in the coming years, sharp recriminations will emerge as the people's sense of betrayal grows with the realization of the precariousness of their situation and the planet's.

The question will be asked — similar to the "who lost China"? question asked when China became a Communist country in 1950 - who lost planet Earth? Why did we not act in time?

See Carbon Crimes Against Humanity, Central Climate

Conundrum, Climate Truth and Reconciliation Commission, Media Omertà

Carbon Retirement See Intentional Grounding

Carbon Scapegoats

A strategy to publicize valid ecological explanations (overpopulation, depletion of resources, pollution, climate change) for such problems as illegal immigration or economic contraction.

This might avert a great deal of unnecessary conflict. Better to make carbon at least a partial scapegoat in the name of reducing political conflict while providing a boost to efforts that address carbon pollution.

Carbon Schizophrenia

The vast gap that exists between national climate targets and actual infrastructure decisions made by those same governments.

The UK, for example, has one of the most aggressive carbon emissions reduction laws in the world, with the 2008 Climate Change Act mandating an 80% reduction in carbon emissions over 1990 by 2050. Yet successive UK governments simultaneously promote policies that encourage greater oil exploration and recovery in the North Sea, in direct conflict with its carbon reduction goals.

The May government also supports a third runway for Heathrow airport, thus supporting expansion of a travel mode that generates sky high, as it were, emissions with little prospect of renewable energies powering flight anytime soon.

This kind of climate schizophrenia is hardly unique to the UK. Most countries in the world are unable for political and economic reasons to swear off carbon-based growth at the same time that they are unwilling to become international pariahs by swearing off vigorous climate change action.

The Trump Administration is attempting to transcend carbon schizophrenia by denying the reality of climate change. Unfortunately the earth is indifferent to such delusions.

See Decarbonization Divide

Carbon Sirens

Ulysses in Homer's *The Odyssey* was so tempted by the allure of the Sirens that he had to order his crew to restrain him from being in contact with them.

Today carbon is our Siren — both enticing and addictive. We have yet to find a way to ignore her charms. Can we find our modern version of Ulysses' army to restrain us from continuing to experience the delights of our very own carbon sirens?

The survival of life on Earth depends upon the answer to this question.

If anyone wants to create a climate army to disarm the Carbon Sirens, please let the planet know as soon as possible.

Carbon Supply-siders

Those who believe that focusing on restricting the supply of fossil fuels is essential to reducing carbon emissions.

Given the widely-accepted analysis prepared by the non-profit group Carbon Tracker that only one-fifth of the proven fossil fuel reserves can be burned to avoid the 3.6º threshold, restricting the supply of carbon is an absolute requirement.

This term is based on supply side economics where the goal is to increase the supply of capital by lowering taxes.

See Carbon Exclusion Zone, Carbon Retirement/Intentional Grounding, Keep It in the Ground, 2 Degree Dogma

---

Carbon Surge

Worldwide carbon emissions increased from about 1.5 billion tons a year in 1950 to almost 10 billion tons a year today. This carbon surge created the conditions for the unprecedented prosperity we enjoy today. It is also directly responsible for the climate crisis that puts our planet at risk.

Carbon Terms

There's no shortage of words to describe the amount and nature of CO_2 emissions. Carbon neutral, carbon positive, carbon negative, zero carbon, zero energy and net zero energy are some of the terms used. There is little standard methodology for these terms.

The ecological architect William McDonough has proposed an alternative vocabulary for carbon emissions using terms such as *living carbon, durable carbon* and *fugitive carbon*. These respectively mean carbon that's part of natural cycles and is supportive of life, carbon that is part of durable materials such as coal, buildings or recyclable materials such as paper, or toxic carbon such as carbon that results from fossil fuel burning, deforestation or industrial agriculture.

http://www.mcdonoughpartners.com/william-mcdonough-offers-new-language-carbon/

Carbon Time Travel

Several of the scenarios developed by the **IPPC** to demonstrate how the earth could remain below a 3.6° increase in temperature required global carbon emissions to peek by 2010 and then decline. The problem, as the climate scientist Kevin Anderson has pointed out, is that these scenarios were published in 2013, and global carbon emissions continued to rise after 2010, thus requiring a kind of carbon time travel to keep temperatures at 3.6°.

http://kevinanderson.info/blog/duality-in-climate-science/

Carbon Tourniquet

Short term, local or individual measures to reduce carbon emissions.

These may be put in place while the more challenging work of transforming our energy and other systems to eliminate carbon emissions proceeds.

Examples of carbon tourniquets would be raising buildings a few feet to minimize flood damage, changing lightbulbs and riding a bike to work.

See Bright Light and Dark Green Environmentalism, Climate Palliation

---

Carbon War Criminal

As climate chaos increases, citizens will start identifying those leaders who knew the grave consequences of not acting on climate chaos and yet chose to thwart action. It is reasonable that the law will evolve to define the kinds of behavior that would classify one as a potential carbon war criminal.

See Carbon Recriminations, Climate Crimes Against Humanity, Climate Truth and Reconciliation Commission

---

Central Climate Conundrum

It may seem impossible to imagine that the most technologically advanced and sophisticated civilization in history could collectively choose, in essence, to destroy itself, but that is what we are now in the process of doing.

We are unprepared to make the necessary changes to our societies to avoid the worst of climate chaos and yet are also unprepared to acknowledge the consequences of not making these changes.

This is the Central Climate Conundrum.

Chronic Climate Management

Climate chaos is more like a chronic disease such as diabetes than an acute infection. We can hope to manage it but are not likely to eliminate it.

Dealing with this will be exceedingly challenging at the individual and societal levels. It will require the persistence of the long-distance runner rather than the short-term stamina of the sprinter. How will a culture where attention spans are exceedingly short handle this?

See Bifocal Behavior, Distraction Dilemma, Never Normal Coalition, No Solutions Coalition, Predictable Unpredictability

Civilizational Preservation Movement/Project

The possible climate related destruction of much of world civilization in the coming decades has led to the creation of institutions and projects whose goal is the preservation of important aspects of our civilization. These efforts include the establishment of seed banks, digital archives and language preservation and restoration projects. In addition, species restoration projects are attempting to use genetic information available from extinct species to restore species.

Examples of these efforts include:

The Svalbard Global Seed Bank. The world's most important seed bank is located under the ice in northern Norway. Sometimes called the 'doomsday vault' and established in 2008, it holds almost a million seeds from all over the world. Its location, design and construction are intended to ensure that the seeds are secure from hazards that could compromise seed integrity.

The Rosetta Project started by the Long Now Foundation catalogues the world's languages. Its uniqueness comes from creating storage platforms that can last for thousands of years.

The Revive and Restore program also created by the Long Now Foundation helps coordinate worldwide efforts to apply genomic technology to preserve endangered species and potentially bring to life extinct species.

Retrieving and preserving ancient knowledge and artifacts is another goal of this movement. One such project verifies, validates and applies ancient systems of weather forecasting to prepare for the changing climate in Peru. Oral histories and story-telling are also important avenues to preserve civilizational knowledge.

https://www.croptrust.org/what-we-do/svalbard-global-seeseed-vault/
http://reviverestore.org
http://rosettaproject.org

Civilizational Torschlusspanik

The German word for "Life is passing you and you haven't done much with your life." It's a recognition by some that the game may soon be over on Earth and it's time to get on with it..

Civimortis

The death of civic life.

Will civimortis result from climate chaos? Or will chaos bring us together and strengthen civic life, as people experience the benefits of 'strength in numbers'? This question is among the most important as we enter a **post normal climate.** The answer will depend on location, severity of climate chaos, resiliency of the population, and indigenous leadership.

Climachondriac

Someone who attributes virtually every event, trend and phenomenon to climate change, whether warranted or not. Perhaps we'll have to censor the Internet (or its successor the **mininet**) to stop folks from devouring the equivalent of WebMD or the Mayo Clinic for climate events.

This is likely to be the fastest growing disease, with nary a doctor equipped to treat it. Where is Woody Allen when you need him?

Climatastrophe

A natural catastrophe largely or solely the result of climate chaos.

Climate Barons

Who will be the barons in the **post normal climate** era, replacing the robber and coal barons of an earlier era?

And where will their riches come from? Will they come from shrewd investments in renewables or from IPO's of companies with new technologies to take carbon out of the air? Will the most sophisticated investors employ elaborate carbon screens to identify companies most dependent on carbon? Or perhaps the barons will be those who simply bet against the **carbon combustion complex** at the right time?

Climate Chronosphere

The ways in which climate effects are distributed in space and time.

Phenomena such as the shifting of the seasons and climatic zones, the greater relative heating occurring at night than during the day, the rapid migration of plants and animals towards colder environments fall under the rubric of the climate chronosphere.

See Biotic Refugees, HiLo, NoHi

Climate Cocooning

An extension of the concept of "cocooning" as coined by the futurist Faith Popcorn in the 1980s. Extreme weather and other dislocations will make travel and even routine outdoor activities more uncomfortable and hazardous. People will remain home watching The Weather Channel and climate cocooning, at least for as long as TV is still around.

Climate Committee to Save The World

Time Magazine had a famous cover in 1999 highlighting what they dubbed the 'committee to save the world'. Saving the world referred to a serious financial crisis that the triumvirate of Larry Summers, Robert Rubin and Alan Greenspan intervened to defuse. These three were very high ranking US government officials.

Many may be awaiting the emergence of a Climate Committee to Save The World made up of equally prominent, skillful and powerful people.

Who these people should be I leave to the reader to determine.

The larger question is what should they do? Visit **Dirty 90** and **Walton Family Climate Fix** for ideas of what could be done.

See Climate Compartmentalization, Integrated Climate Planning and Management

Climate Compositions

The creation of original musical compositions designed to represent, model or evoke climate change.

See (and hear):

http://www.treehugger.com/climate-change/listen-133-years-climate-change-in-one-evocative-song.html
https://www.outsideonline.com/2109116/art-turning-climate-change-science-music

Climate Contraception

As climate extremes worsen and life becomes more dangerous and onerous with no end in sight, the birthrate is likely to plummet as families make climate contraception a central practice in their lives.

The birthrate is also likely to decline even for families who desire children. Research shows that the number of babies born in the US nine months after a hot day was over 1,000 fewer than expected. Whether that's from less sex or biological changes in fertility is not known.

http://www.reuters.com/article/us-usa-climate-birthrate-idUSKCN0SZ0XM20151110

Climate Denial Deprogramming

Giving the tenacity with which climate deniers appear to hold their beliefs, changing a denier's mind may be a painful and protracted process.

While it is unlikely that climate deprograming will literally be undertaken, incorporating this idea into discussions about climate denial dramatizes the degree to which denial should be viewed as a cult like activity divorced from any connection with reality.

See Climate Denial Spectrum, Denial Triad, Denial Two Step

Climate Denial Spectrum or the Denial Two Step

The explanations that climate deniers give for inaction on climate change are ever shifting. Each position shift occurred throughout the denial community at about the same time, demonstrating remarkable intellectual dexterity or more likely remarkable political machine like discipline.

Initially, deniers claimed that global heating wasn't happening, then begrudgingly they acknowledged it may be happening, but it's the result of natural causes. When that became a hard argument to sustain they said ok, maybe humans do have something to do with it but climate change is good for us. As that argument collapsed, the next shift was that maybe it's not good for us, but we shouldn't do anything about it as it will hurt us economically. They then segued to arguing that doing something won't help anyway.

Next came the argument that we shouldn't do anything because the Chinese aren't doing their fair share. After the US and the Chinese jointly announced their ratification of the Paris Climate Agreement in September 2016 that argument evaporated.

And then, perhaps their most laughable argument was made. The coordinated refrain from all sectors was "we're not scientists so we don't know what the hell is going on."

At the time of this writing, our new President is back to scapegoating China, suggesting that climate change is a hoax perpetrated by those perfidious Chinese. (Perhaps not surprisingly he has used climate change as an argument for building a 13-foot wall – no, not *that* one - to protect his Irish golf course.)

The climate denial spectrum is constantly changing as one bogus argument after another gets knocked down by inconvenient truths. Anyone taking bets on what the next desperate assertion will be?

In the face of immense suffering for so many it is truly tragic that this farce continues unabated. It's about time for the denial two step to be replaced by another two steps — one called truth and one called reason.

See Denial Triad

https://www.theguardian.com/environment/climate-consensus-97-per-cent/2016/may/26/donald-trump-wants-to-build-a-wall-to-save-his-golf-course-from-global-warming

---

Climate Disobedience

It is remarkable how little climate related civil disobedience is occurring. Some civil disobedience has occurred around the Keystone and Dakota Access Pipelines and in other scattered protests.

Perhaps it's because there has yet to emerge a charismatic leader to rally Americans and others around the world to tackle climate chaos through massive civil disobedience.

No Martin Luther King Jr, no Mandela, no Gandhi, no Lech Welesa or Vaclav Havel. Nor has there been a defining event — a Stonewall riot, or the March on Washington or Pearl Harbor...yet.

Or perhaps we don't care enough or feel too powerless to put our bodies on the line.

Climate Exceptionalism

One could argue that rallying the world to mobilize against climate change would be the purest form of American exceptionalism.

Yet American climate exceptionalism has taken the form of climate denial, or at least it has in the statements and actions of one of our political parties.

We stand as the country now most exceptional in our stand against reason and planetary survival. Which I guess is another form of exceptionalism, however dismal, dispiriting and irresponsible that it may be.

Climate Fatalism

The state of mind that many will experience as the reality of the scale and depth of climate chaos becomes evident.

See Climadelic Therapy, Climate Fatigue, Oh Shit Moment

Climate Fatigue

The physical and emotional exhaustion that people focused on climate chaos will experience given the immensity of the problem and the apparent difficulty if not hopelessness in ameliorating or overcoming it. Often climate fatigue will lead to **climate fatalism**.

Climate Glocalism

Unlike most other environmental issues, addressing climate chaos requires both local and global (or 'glocal') action. Local, as all carbon pollution is generated at a particular stationary or mobile location — whether it be a power plant, a car, a forest or a melting ice cap. Yet local action is clearly insufficient — indeed futile — without international action. This is because carbon emitted anywhere winds up everywhere.

Thus, there is little inherent environmental or economic incentive for anyone to reduce carbon emissions, as the benefits will accrue to all seven plus billion of us. Shared responsibility, empathy, altruism and the promise of a more livable world must be the prime motivators for action.

See Bifocal Behavior

Climate Horizons

Many governments, corporations and political leaders typically choose a time horizon of 2040 or 2050 when they set targets for carbon emissions reductions.

While essential to focus on the long term, projections and targets that go beyond a decade or two have little ability to motivate people to act. In addition, climate science demonstrates the much greater value of emissions reductions that occur sooner rather than later.

Announcing ambitious climate targets for 25 to 35 years from now gives the appearance that one is a strong **climate hawk** without having to make the hard choices and trade-offs that shorter term aggressive actions would require.

The interested climate hawk should demand that our leaders focus on setting ambitious targets with climate horizons of 10 to 15 years as well as longer term horizon years to help insure that action is taken now and not put off until 2039 or 2049.

See Placebo Plan, Procrastination Penalty

https://theconversation.com/2050-climate-targets-nations-are-playing-the-long-game-in-fighting-global-warming-69334

Climate Humility

The state of mind that acknowledges that our collective knowledge of climate chaos related causes, consequences and remedies is quite a bit less than what is needed to fully understand and intelligently respond to the challenge. Or more simply, it's damn hard to predict what climate chaos will look like, how quickly it will worsen and exactly what to do about it.

Climate Manhattan Project

The US should initiate an emergency climate research and development effort on the scale of the WW2 Manhattan atomic bomb project.

The US is vastly underinvesting in climate related research and development, as compared to the need. Columbia Professor Jeffrey Sachs makes a compelling case that we need to focus national efforts on what he calls 'directed technological change'. Such an effort would address the need to develop climate related technologies on an urgent, coordinated and well-funded basis.

See Climate Plan A and B, Eemergency

http://www.bostonglobe.com/opinion/2016/11/27/big-innovations-require-big-investment/d6nm8c4mVzo2NMSHQDFK7I/story.html#comments

Climate Missionaries See International Climate Corps

Climate Palliation

Those activities that provide a measure of relief from the consequences of climate chaos without addressing the urgent, sweeping and often radical actions that must be taken to significantly reduce the effects of climate chaos.

Raising street elevations in a city for example would be climate palliation as opposed to a permanent prohibition of development in flood prone areas. The use of **climataceuticals** is also an example of climate palliation.

At the personal level, climate palliation can provide a level of physical and emotional support to those at risk from climate change that is less intensive than support provided by planetary hospice services.

See Carbon Tourniquet

Climate Plan A and B

What science unambiguously demands is a full WW2 scale immediate worldwide mobilization to decarbonize in a generation or less, a **One Generation Challenge**. As Bill McKibben says, "That is Plan A".

If that does not happen, (and optimism that Plan A will happen is in short supply), a Plan B must result in a massive worldwide adaptation campaign, a defensive ever-changing effort at building physical resilience into our cities, towns and countryside, and emotional resilience into our bodies, our hearts and our minds.

See Adaptation/Mitigation, Anticipatory Reparations, Climate Manhattan Project, Readaptation

---

Climate Shapeshifting

Climate chaos is already resulting in remarkably swift evolutionary changes on the part of some life forms in an attempt to survive increasingly warm conditions.

Perhaps the most common evolutionary adaptation will be the shrinking of body mass, as a smaller body results in proportionately more skin area and thus greater cooling as per these examples:

Certain species of salamanders have decreased in size by eight percent in 50 years. A recent study projected that fish could shrink by up to 24% by 2050. The zebra finch changes its tune to produce smaller chicks that can better survive, as described in **changing your tune about climate change**.

http://www.nature.com/nclimate/journal/v3/n3/full/nclimate1691.html

Climate Singularity

The belief that continued progress in artificial intelligence will lead in a matter of a few decades to the 'singularity' where machine or artificial intelligence becomes superior to human intelligence. Soon thereafter, man can essentially live digitally. The need for a habitable climate then becomes optional. So, we hope, or then again, do we?

Climate Trigger Warnings

Notifying people prior to potential climate disruptions so they can properly prepare for the unpleasant events to come.

These trigger warnings, unlike the ones issued on many campuses, may not allow one the option to skip the event in question that triggered the warnings.

Climatism

An extreme preoccupation with and focus on climate chaos. One would expect that someone might embrace climatism soon after experiencing an **oh shit moment**.

See Climachondriac

Climatographers

Professional or citizen photographers documenting the effects of climate chaos throughout the world.

Climatographers are playing a critical role in conveying the effects of climate chaos to local and international audiences. Their courage in climbing on the last bit of a melting glacier, or following a flooded out and bewildered family in a hundred different countries, or staring into the torso of a bloated and misshapen sea mammal that lost its life to a **marine heat wave** is to be commended.

Climatrarian

An individual whose beliefs about climate chaos reflect an alternative, non-mainstream or unpopular perspective. Perhaps it's the **climate doomers**, or the brightest of the **brightsiders**, the **climate singularity** folks, or the **marsitects.** And let's not even mention the climate deniers.

Clitanic Era

So much of climate chaos is hidden from view.

We can't see ocean acidification nor the methane about to emerge from permafrost. We can't see the fugitive emissions that warm the atmosphere nor changes in plant respiration. We can't see the marine heat waves that bake ocean waters, nor can we see most of the plants and animals migrating daily in search of cooler climates and the secure life they no longer have. And we can't see most of the hundreds of thousands of people who are estimated to die from the impacts of climate change each year.

Like the iceberg that sank the Titanic, we can see a small fraction of the changes brought about by climate chaos if we look for them. The melting glaciers around the world, the sunny day floods in Miami Beach, the jellyfish and algae blooms keeping us out of the water on our summer holidays are a few of the not so hidden effects of climate chaos.

So yes, welcome to the Clitanic Era, a subset of the **Anthropocene**. Welcome to what might be the last act in an incredibly short but spectacular run by Homo sapiens. A run that might end not just with our species tattered and decimated, but with the vast majority of our brethren fellow travelers swept away not by comets or volcanoes or sunspots, but by us.

And let us hope there will be someone around to make the films and write the books about the Clitanic, just as there were lots of us around to make the films and write the books about the Titanic.

Deadlihoods

Neighborhoods likely to receive disproportionately negative effects of climate chaos, in the form of flooding, water shortages, disease, or dangerous temperature levels.

Death Events

The frequency and magnitude of large scale events leading to death of humans or other species is likely to dramatically increase as a result of climate chaos. Just as weather forecasters report on weather events so the news will increasingly consist of the recitation of climate related death events.

See Carbicide

Decarbonization Divide

To have a reasonable chance to meet the 3.6º maximum temperature increase deemed relatively safe (See **2° dogma**), decarbonization must occur very rapidly starting NOW. The annual reduction in the amount of carbon that is released into the atmosphere must be on the order of 6 to 10% or more.

In practice, CO_2 has never declined as growth has occurred. (Emissions were flat over the past three years while world GDP grew modestly.) The decarbonization divide is the difference between the 6 to 10% or more of annual decarbonization required and the percentage that historically can be expected (close to zero).

The smaller the decarbonization divide, the higher the degree of stranded assets, as those assets, primarily fossil fuels, will be abandoned or written down much faster. Too great a pace of decarbonization also runs a risk of a severe economic turndown, which in turn will reduce **global dimming** through less carbon and other pollutants being burned. This will paradoxically accelerate global heating. So, the result of decarbonizing as necessary to prevent catastrophic climate chaos may well be a severe economic crash, trillions of dollars of stranded assets, and accelerated global heating.

And that's the good news. The bad news could be that **hyperfeedbacks** come into play and create havoc before the world can sufficiently reduce carbon emissions.

See Stranded Studies

Deniers Triad

Climate denial does not seem to have diminished in the United States despite the increasing real time experience of climate change. The Six Americas study done by Yale and George Mason identified 11% of the population as dismissing climate change in 2007 rising to 13% in 2015.

It appears that much of the impetus for climate denial comes from one or more of these three factors:

- Religion
- Ideology
- Economy

Religious deniers tend to be those with fundamentalist views. Some believe that if God wishes for the climate to change, then so be it, and we must not interfere. Related to that attitude is the belief that end times are near and climate change is the manifestation of these end times. Others feel that God would not allow devastating climate chaos to occur and therefore it cannot be occurring.

Most religious deniers associate climate change with liberal countercultural movements, which are anathema to most of them.

(Contrast these beliefs with Pope Francis's perspective that God will judge a person on whether they cared for the Earth.)

Another group — the ideologues — resists climate chaos based not so much on the reality of climate chaos, but on the remedies. Remedies that will require centralized government and vigorous intervention in the economy and in people's lives. Having to give up one's passion for small government for climate action is not a tradeoff many ideologues will make.

The third group consists of the real world 'pocketbook deniers' who would lose income, employment, and power as fossil fuels become restricted. It's quite straightforward for these folks and institutions to fight against the reality of climate change.

Many will have views and motivations that place them in more than one of these categories.

The denial triad is a formidable force that continues to resist reality despite unassailable science and thousands of examples of physical evidence all around us.

See Climate Attitudes, Denial Spectrum Disorder, Denial Two Step

---

Desert Riviera

More than just a name for a hotel in California, Desert Riviera could become the new name for the French Riviera. Or so suggests a 2016 French study that concluded that much of the Mediterranean could become desert even with warming limited to 3.6°.

https://insideclimatenews.org/news/27102016/global-warming-mediterranean-region-desertification-drought-climate-change

---

Desperation Dividend

The benefit that comes from recognizing how desperate a situation may be.

Several *CVF* entries address the endemic complacency that many have towards climate action. Perhaps only a catastrophic storm, climate related epidemic disease, financial collapse or other major and sustained climate related calamity can lead to a level of desperation that results in climate action. The intense reaction to the climate denial and policy agenda of the new President appears to be leading to at least a small desperation dividend.

See Procrastination Penalty

DETS Crisis - Debt, Energy, Temperature, Species

Never has the world experienced such unprecedented levels of debt (over $200 trillion or three times world GDP), energy issues (peak oil, low price of oil and high costs of exploration, pressure to keep oil in the ground), temperatures (the highest temperatures in recorded history occurred in 2016), and species loss (the extinction rate is estimated at 1,000 to 10,000 times higher than the background rate). The potentially explosive and difficult to predict interaction between these four phenomena deserves close attention.

See Peak Oil Debt, 2° Dogma

http://www.bloomberg.com/news/articles/2016-02-22/the-world-s-debt-is-alarmingly-high-but-is-it-contagious
http://www.biologicaldiversity.org/programs/biodiversity/elements_of_biodiversity/extinction_crisis/

Dirty 90

Ninety companies extracted almost 2/3 of the total carbon-based fuels burned since 1751 according to a 2013 report by Richard Heede. 83 of the 90 produce fossil fuels and 7 produce cement. The largest producers are the state owned or affiliated companies in Russia and China.

A thought experiment:

A billionaire who understands that his money and his family's lives will be worth nothing if climate chaos continues unabated invites the CEO's and CFO's of each of these 90 companies to a week-long retreat at a super luxury resort. Awaiting these guests is a group of the world's leading climate scientists, economists, and financiers. Their goal is to demonstrate how the Dirty 90 can transition away from fossil fuel production in 10 years in ways that preserve their companies while preserving the world.

Who might be that billionaire? Will the Dirty 90 now kick the fossil fuel habit and become known as the Clean 90? Could they become the **Climate Committee to Save the World?**

See Waltons to the Rescue

http://thebulletin.org/just-90-companies-are-accountable-more-60-percent-greenhouse-gases10080

Disaster Denialism

Climate change is not the only disaster where large numbers of people ignore reality. The aftermath of the 1906 San Francisco earthquake is instructive and is eerily similar to the actions of the Federal government and some states.

> The San Francisco Real Estate Board passed a resolution to rebrand the "San Francisco Earthquake" as the "San Francisco Fire." The California State Board of Trade (the future Chamber of Commerce) would reference the San Francisco "disaster" or "catastrophe" but never uttered the dreaded "e-word": earthquake.

There is not one monument to the earthquake in San Francisco, and in 2004 Daly City, Ca, the epicenter of the quake, turned down a request to erect a monument to the earthquake.

In yet another disaster denial, remember the movie *Jaws 2*? The town leaders chose to deny the evidence of a shark in order to keep the beaches open and the tourists in town.

http://www.sfgate.com/bayarea/article/DALY-CITY-Officials-unmoved-by-quake-notoriety-2768399.php

Disaster Emissions

CO_2 emissions that result from natural disasters that in turn result from CO_2 emissions.

Prime examples are fires that result from drought, dangerously high heat caused by climate chaos or CO_2 induced heat that briefly transformed the Amazon basin into a source of carbon rather than a carbon sink in 2005 and again in 2010.

See Carbon Clogs, Man Made Natural Disasters, Sources and Sinks

Distraction Dilemma

Achieving the long-term carbon reductions required to minimize climate chaos requires intense political, economic and behavioral focus. Yet maintaining this focus is immeasurably complicated by the endless distractions in our individual and collective lives in modern societies.

Collective distractions such as terrorism, political controversies, celebrity obsession, Monday night football, and plane crashes half-way across the earth often dominate our lives. Keeping your boss happy, waxing your skis and finishing the book in time for your Oprah's book club meeting are some not uncommon personal distractions.

Terrorism is perhaps the most pervasive distraction. How unfortunate, as it monopolizes the airwaves, scares people well beyond the objective risk, and threatens to elect candidates who overhype the terror threat while denying the climate threat. (Does that sound like someone who was just elected President?)

The noticeable reduction in attention span brought on by electronic devices is no help either. What is particularly alarming about our reliance on devices is the increasing evidence that those devices in some very real way rewire our brains, perhaps making it much more difficult to focus on climate chaos *even if we want to*.

What distracts us?

What price do we individually and collectively pay for focusing on these distractions?

How can we transform the distraction dilemma into the attention attitude?

See Frozen Chicken Syndrome, Kardashian Climate Ratio

http://www.cbc.ca/life/wellness/how-smartphones-are-rewiring-our-brains-1.4036816

Dodgy Climate Pledges See Climate Horizons, Placebo Plan

EarthPrint

The sum of all impacts — positive and negative — that humans have on the earth's ecosystem.

Analogous to a carbon footprint for a person, process or collective enterprise, an earthprint is calculated for the entire planet. Our earth-print needs to shrink and fast if we are to have a chance of maintaining a habitable planet.

See Natural Balance Sheet, Planetary Boundaries

Eat Your Carbon Spinach

A common communications approach to educating audiences about climate chaos that emphasizes threats, hardships and difficult behavioral changes.

Even with the growing popularity of veganism an eat your carbon spinach strategy is perhaps not the optimal way to motivate most people.

Ecocholia

A kind of melancholia that results from experiencing the climate chaos induced fate of the earth.

Who would be immune from ecocholia once an **oh shit moment** is experienced?

See Solastalgia

Ecostasis

A time when the earth's ecological systems are in rough equilib-rium.

As we enter the **post normal climate** era, the likelihood of significant ecostasis is minimal. Many will look back with **solastalgia** at the relative ecostasis of the late Holocene early **Anthropocene** era.

See Punctuated Disequilibrium

EEmergency

No phenomenon in modern history except for nuclear war (or perhaps an attack from outer space) is likely to create as much ongoing sustained damage as climate chaos. Billions of lives are beginning to be affected in a sustained and essentially never ending manner.

The inadequacy of words to describe this experience requires a new word for this emergency. That word is eemergency.

What would an eemergency look like? Let's start with the recognition that the threat is so imminent, so potentially catastrophic and so pervasive that immediate action is required. Action is required on a scale equivalent to if not greater than our mobilization for WW2.

All actions follow from this premise — a premise currently shared by fewer than one in five Americans according to the Six Americans survey. Climate mobilization was articulated in the 2016 Democratic Party platform. "We are committed to a national mobilization, and to leading a global effort to mobilize nations to address [the climate crisis] on a scale not seen since World War II."

However, this mobilization was never emphasized or even mentioned by the Democratic nominee for President in 2016.

theclimatemobilization.com, a small non-profit has prepared an ambitious and detailed "Victory Plan" for eemergency climate mobilization. Have a look.

See One Generation Challenge, Six Americas

http://thebulletin.org/what-it-really-means-fight-climate-change-war9921

Fire pollution

The effects of widespread climate chaos induced fierce forest fires extend well beyond the areas directly impacted. What could be called fire pollution can spread for hundreds of miles.

After fires in Quebec, Canada in 2002 there was up to a 30-fold increase in harmful airborne fine particle concentrations contained in an atmospheric dome that extended 1000 miles to Baltimore, MD, according to the National Oceanic and Atmospheric Administration. And many of California's fires adversely affect air pollution and visibility in states located hundreds of miles away in the interior west.

Much of the air pollution from these wildfires takes the form of PM 2.5 particulate particles, which are very small particles that can burrow deep within lung tissue causing serious lung and heart disease. Remember that the next time you shrug off the news of a serious wildfire hundreds of miles away.

First Flee/rs

The phenomenon of investors and homeowners moving from their communities after anticipating which neighborhoods, cities or regions to move to based on projected climate impacts.

Generally the areas being fled to will be inland, more northern than southern and at a higher rather than lower altitude. Early adopters (and adapters) will also tend to move to areas that have governments that accept climate change and have taken aggressive **adaptation** actions.

See HiNo

Fossil Fuel Preserves See Carbon Exclusion Zones

Frequent Flyer Fee

A proposal to tax flyers based on the number of flights taken in a year. The more one flies, the higher the fee would be for each flight. This would reduce airline demand by targeting those who can afford flying and thus help reduce carbon emissions from the airline industry.

The British group afreeride.org has proposed such a levy to raise money for alternatives to flying and to stop the construction of new runways at airports in London.

Friedman Climate Unit

Adapted from the 'Friedman unit', a term created by the blogger Atrios to describe a frequent comment that the New York Times columnist Thomas Friedman made about the prognosis for the Iraq war. He was said to have indicated that the next six months would be critical in terms of turning around the war and prevailing. It seems however he would make this observation some 14 times over the course of several years.

Similarly, many argue — using a 'Friedman climate unit' — that we have perhaps 10 years to begin a serious effort to significantly reduce our carbon emissions to prevent calamity. Unfortunately, these proclamations were — and are — made year after year even as the time to act with urgency is right now.

See Procrastination Penalty

https://en.wikipedia.org/wiki/Friedman_Unit

Friends of the Enemies of the Earth

An organization that is the political opposite of Friends of the Earth.

While not an actual organization — yet — the Koch brothers, oil, gas and coal companies and their friends (or shall we say employees) in Congress and in state legislatures could well be charter members of this organization.

Frozen Chicken Syndrome

Named after the practice of shooting chickens from guns to test the strength of airplane windshields.

According to the likely urban legend a frozen, rather than a fresh, chicken was accidentally shot from the gun and smashed through a train windshield window and into the next compartment. (British Rail supposedly borrowed the technique — if not the chickens — from the American plane testers.)

Whether or not the story is true the **frozen chicken syndrome** is meant to convey the fact that stuff happens, things go wrong and snafus (**S**ituation **N**ormal **A**ll **F**ucked **U**p) occur.

We have to transform our energy and agricultural systems, our dietary practices, as well as our transport, housing and urban sectors in a very few years to have a chance of preserving some semblance of a livable planet. To do this will require extraordinary commitment and unity, precision planning and almost flawless execution worldwide. Yet the frozen chicken syndrome will be lurking behind every decision and every action, reminding us of all that can — and perhaps will — go wrong along the way to a zero-carbon world.

See One Generation Challenge

Gigatastrophe

An unimaginably large **climatastrophe** of the kind that is likely to be a frequent occurrence as climate chaos intensifies.

Great Gamble

Can the 7+ billion humans on earth reduce greenhouse gas and related emissions by a large enough quantity in a short enough time to avoid catastrophic climate chaos without greatly inconveniencing ourselves?

The Great Gamble is our collective calculation that we can get away with minimal efforts to address climate change.

As of this writing, it appears we may be losing The Great Gamble. But the window hasn't quite closed, so place your wagers.

See Central Climate Conundrum

Gretzky Climate Strategy

"Skate to where the puck is going not to where it was" or in climate terms anticipate the range of **climate positive feedbacks** and CO_2 levels and plan your interventions to address these potential conditions. While the Gretzky climate strategy is one of the more overused aphorisms there is a certain utility in applying it to climate chaos.

One example where it could have been usefully applied is the paucity of climate models simulating what life on earth will be like with a 6° or 7° (should life still exist at that level of warmth) or greater increase in temperature. Many scientists chose not to model the consequences of that high a temperature increase as they understandably assumed that the world community would effectively intervene to prevent that level of global scorching. Sadly, that does not appear to be the case.

Guilt Per Gallon See Micro Guilt

Guns vs Carbon

The US spends about 28 times as much on traditional defense as on climate change defense according to a 2016 report from the Institute for Policy Studies. Since climate chaos is the ultimate existential threat to our country and the world, shouldn't that ratio be reversed or at least equalized. And this was under the Obama administration!

See CDARPA

http://www.ips-dc.org/wp-content/uploads/2016/09/CvsC-Report-1.pdf

Heat Harm

Research is uncovering an increasing number of ways that global scorching induced high heat harms us and our surroundings:

- Heat stroke, suicides and other deaths increase.
- Kidney failure due to dehydration from high temperatures is estimated to have caused some 20,000 sugar field laborers to die in Central America over the past 20 years.
- Productivity decreases by about 2% for every degree of temperature above about 80º.
- Births decrease nine months after temperatures rise above 80°.
- Student test scores decline.
- Food and medicine spoil.
- Crime increases.
- Roads, bridges, train tracks and electrical signals buckle, endangering travelers.

See Climataceuticals, Climate Contraception, Death Events, Natick Climate Center

https://www.theguardian.com/news/2016/jun/20/claimate-change-bad-kidneys-weatherwatch
https://www.ncbi.nlm.nih.gov/pubmed/27822625

Heat Islands

Localized areas of higher temperatures typically found in urban areas as a result of dark surfaces, industrial activity and heat from human beings and animals.

The planting of trees and the painting or replacing of dark surfaces with light surfaces and green or white roofs can reduce heat islands. These measures also contribute to comfort for residents and lower energy use.

While some advocate for large scale programs to paint roofs to increase the urban **albedo,** research suggests that the climate benefits of white roofs may not be particularly large.

Roads are also a significant contributor to heat islands. Asphalt is a particularly unhelpful material given its derivation from oil and its black non-reflective surface. Replacing asphalt however with the lighter color of concrete would result in use of a product that is among the single largest generators of carbon emissions, demonstrating yet another example of the **carbon maze**. Maybe it's time for solar roadways, as the linked article below suggests.

To add another zig (or zag) to the carbon maze a late 2016 study concluded that the cement in concrete reabsorbs some, but not all the carbon that was emitted during its manufacture over a lifetime of use. So, cement is still bad for the climate, but not quite as bad. Whether it is better or worse than asphalt remains to be determined.

See White Only Laws

https://www.eurekalert.org/pub_releases/2016-11/uoc--cjf111516.php
http://www.bloomberg.com/news/articles/2016-11-24/solar-panel-roads-to-be-built-across-four-continents-next-year

Help Kelp!

Studies have shown that the underwater kelp forests off the coast of Australia and elsewhere are rapidly shrinking because of warmer weather and hotter ocean temperatures. It appears that the warmer ocean temperatures attract more and hungrier fish to the kelp, eating them faster than they can regrow.

Those who advocate for the cultivation of large-scale seaweed farms to draw down CO_2 must wrestle with the challenges posed by the destruction of kelp and other seaweed by climate chaos.

See Climate Maze, Coral Bleaching, Reef Races, Seaweed Farms

https://www.theguardian.com/environment/2016/jul/07/australias-vast-kelp-forests-devastated-by-marine-heatwave-study-reveals
http://www.science.unsw.edu.au/news/study-underwater-video-reveals-culprits-behind-disappearance-nsw-kelp-forests

HiLo

Climate chaos results in relatively greater increases in night temperatures than in daytime temperatures. Many more **Hi**ghest recorded night time **Lo**w temperatures have been broken than daytime high temperatures.

Scientists believe this phenomenon may result from a thinner boundary layer of air at night than during the day. Higher night temperatures can be more dangerous for some than higher daytime temperatures.

Human Tropophytes

Plants that are adapted to thrive in environments with both heavy rains and drought are known as tropophytes.

Can we transform ourselves to be as adaptable as these plants?

Hyperfeedback

Multiple **climate positive feedbacks** interacting in largely unknown but reinforcing ways.

Climate positive feedback mechanisms are perhaps the most feared causes and consequences of climate chaos. Very little is known about the tipping points for each positive feedback, the magnitude of the consequences should a tipping point be reached, and the interaction between these feedbacks.

And because so little is known about these interactions they are generally not included in climate models, potentially vastly underestimating damage from global temperature increases.

An example of a hyperfeedback would be a reduction in the **albedo** in the Arctic as a result of global heating leading to melting ice which leads to more warming which results in the thawing of permafrost (**permadeath**) which in turn liberates methane, which leads to further warming which results in more wildfires leading to more warming and so on. Further intensifying the hyperfeedbacks could be **land carbon-climate feedbacks** or carbon releases from the soil due to increased microbial respiration rates.

Information Underload

While there is much discussion of information overload, there is little discussion of the opposite phenomena, too little information.

The climate field may be particularly subject to information underload. The media mostly ignore climate change, many schools do not integrate climate change into the curriculum and governments like those in Florida, Wisconsin and North Carolina attempt to restrict even the use of the term 'climate change'. The House of Representatives Budget Committee even attempted to prohibit the CIA and the Defense Department from studying climate change. And the real estate industry fights over climate change related disclosures on property in flood zones.

These and many more are examples of climate related information underload. Information underload can be a much more dangerous phenomenon than information overload.

See Media Omertà

https://www.thenation.com/article/what-we-talk-about-when-we-dont-talk-about-climate-change/
http://www.nytimes.com/2016/11/24/science/global-warming-coastal-real-estate.html

---

Intentional Grounding

The **Keep It in the Ground** movement seeks to sharply limit the supply of fossil fuels by using all necessary means to intentionally ground fossil fuels to keep them from being burnt.

To have a chance of not exceeding 3.6º, the world must find a way to avoid burning at least 80% of the proven oil reserves in the ground. At our current rate of carbon combustion, the world will reach the limit of its **carbon budget** in just a few years.

See Carbon Budget, Carbon Retirement, Carbon Supply-siders, Keep It in the Ground, 2° Dogma

---

Jellylands See Methane Craters

Junk Assets See Legacy Infrastructure

Leapfroggery

Technology that for reasons of timing, infrastructure conditions or economics is deployed while skipping the prevailing or **legacy infrastructure** or technology.

The classic example of leapfroggery is bypassing the installation of land line telephones in less developed countries for wireless telecommunications. And of most relevance to climate change is the installation of wind, solar and other renewal energy technologies while leapfrogging over fossil fuel energy generating infrastructure.

The importance of leapfroggery lies in its ability to bypass high carbon infrastructure for renewable technologies that provide modern services at affordable costs.

Legacy Emissions

Carbon emissions from earlier times linger in the atmosphere for many years and cumulatively contribute more to climate chaos than do current carbon emissions.

Some have called for corporations and governments to be accountable for these emissions.

It would be quite illuminating if a way to identify and label carbon atoms that result from legacy emissions could be developed. This advance could make clear the connection between our atmosphere now and human activities spanning the globe over the past couple hundred years. The emissions from the manufacture of muskets for use in the Revolutionary War for example are largely still in the air.

See Carbon Dye, Carbon Fingerprint/Signature, Climate Debt

Legacy Infrastructure

Parts of the world's infrastructure will need to be retired or significantly retrofitted to reduce carbon emissions. While some infrastructure will require little more than energy and water efficiency retrofits, other infrastructure is largely incompatible with a **zero-carbon** emission life cycle and may need to be retired or replaced. Certainly, most coal and natural gas based power plants may be considered as legacy infrastructure.

Legacy infrastructure also covers infrastructure that could be ruined by climate chaos, such as infrastructure in flood prone areas, in extreme drought areas, or in areas subject to major and frequent wildfires.

Airports may become legacy infrastructure unless ways are found to power flight with low carbon sources.

Also known as *Junk Assets* or *Legastructure*

Legacy Thoughts

Perhaps the most critical legacy is the thoughts, beliefs and values that are not in tune with a **post normal climate** world. In the climate chaos context, examples of legacy thoughts would include the religiously derived belief that humans are incapable of changing the climate, or that the climate is not changing, or that technological innovation will guarantee a 'solution' (**technohubris**) to climate chaos.

Legastructure See Legacy Infrastructure

Locavestments

Investments made in local companies or ventures.

Locavestments are likely to grow due to restrictions on travel, particularly air travel, and the increasing localization of food and other goods and services that improve community resiliency.

Low Carb-On Diet

A diet that minimizes carbon emissions and carbohydrates.

Studies point to a diet lighter in meat and focused on fruits and vegetables as generating fewer carbon emissions. Some research has shown that grass fed beef and better range management practices may also reduce carbon emissions.

Foods like grains that are high in carbohydrates and low in carbon emissions present a dilemma for the low carb-on folks.

Macrosilience

Resilience at the largest scales, as at the oceanic, national and international levels. A quality particularly important in dealing with climate change.

Maldives Dive

Former Maldives President Mohammed Nasheed became an international climate hero in 2009 when he announced that his country would become climate neutral in 10 years. He organized a cabinet meeting underwater to dramatize the mortal threat to the country from rising sea levels, and made a stirring speech at the Copenhagen COP 15 in 2009 urging action to reduce carbon emissions to 350 parts per million.

In 2012 Rasheed was deposed in a military coup and imprisoned. The next administration then prepared an **INDC** that called for a reduction in emissions of only 10% by 2030.

This Maldives Dive from potential **carbon neutrality** to 10% illustrates one of the many potential obstacles — in this case political instability — that countries may face in meeting their carbon emission targets.

See Carbon Maze, Distraction Dilemma, Frozen Chicken Syndrome

http://www.gq.com/story/maldive-islands-global-warming-male?printable=true

Marine Sensory Disorientation Syndrome

Some fish exhibit disorientation and confusion as a result of ocean acidification and higher CO_2 levels. Changes in their ability to hear, see and smell appear to be affecting their ability to navigate, with unknown consequences for marine life and health.

Let us hope that land based life will not be subject to a similar sensory disorientation syndrome due to global heating.

http://www.iflscience.com/plants-and-animals/young-fish-cant-navigate-acidifying-oceans/

https://www.eurekalert.org/pub_releases/2016-10/uoe-cci102116.php

Media Omertà

A media omertà — as in the Mafia code of silence — appears to be in place throughout mainstream television. This omertà seems to forbid no more than a passing mention of climate chaos.

If the reader is skeptical of this assertion, perhaps a few statistics from the non-profit Media Matters will raise an eyebrow or two. In 2015, the four networks broadcast a total of 146 minutes (not hours) of climate chaos stories during the entire year. 39 of those minutes were on the one network that denies climate chaos even exists.

The paucity of coverage came in a year that was notable for major climate chaos news: The Paris Climate Agreement, the US Clean Energy Plan, Pope Francis' encyclical on climate chaos and ecology Laudato Si, the Keystone Pipeline, the unprecedented increase in average temperature and many extreme weather events.

Similar minimal coverage occurred in 2014 and in previous years. In the four presidential and vice-presidential debates in 2016 *not one question about climate change was asked*. Amazingly, climate coverage in 2016 dropped by 66% over the coverage in 2015.

Beyond simply the number of minutes of coverage is the content of this coverage. Very little of the coverage emphasizes the dire situation we're in, and the many near intractable obstacles that we face in successfully taking effective action. Did anyone see a story about the **carbon budget** or ocean acidity increases or how the carbon emission pledges made in Paris fall short of what is needed for a habitable planet? I didn't think so.

When historians look back at how we collectively shrugged our shoulders at climate chaos for so many years, near the top of the list of climate villains will be the mainstream media. Their collective media omertà helps create and enforce a passivity and indifference to climate chaos that drains the populace of any motivation to act.

See Carbon Crimes Against Humanity, Social Proof

http://mediamatters.org/research/2016/03/07/study-how-broadcast-networks-covered-climate-ch/208881 https://www.mediamatters.org/research/2017/03/23/how-broadcast-networks-covered-climate-change-2016/215718

Micro Guilt

The small nagging guilt that people are beginning to experience when they get into their car, buy a consumer good or otherwise take an action that releases carbon into the atmosphere. For some, their **guilt per gallon** may be more important than their miles per gallon.

Micro Mitigation

Six (seven billion less infants and institutionalized people) billion people on earth each make many decisions every day that could result in an increase in CO_2 emissions. If each person is assumed to take 10 actions a day that affect CO_2 (walk rather than drive, eat greens rather than a beefsteak, buy gifts for your kids or not, and so forth) that is 10 x 6 billion or 60 billion individual decisions each day or 22 *trillion* decisions annually, most all of which have the potential to micro mitigate CO_2.

How can we best influence the majority of these 22 trillion decisions to avoid climatastrophe within a decade or two?

Mininet

The Internet is likely to shrink to the size of a mininet due to its prodigious consumption of energy and as some of its servers, cables and even users are wiped out due to climate chaos.

Motion Generated Energy (MGE)

Power produced as a result of human or naturally generated motion.

Imagine if every time we walked or rode a bike, climbed steps or jumped our motion was harnessed to generate energy that powered our home or devices.

One example of MGE is the 2016 installation of a specially tiled sidewalk in Washington, DC that generates power that lights a nearby park every time someone walks on it.

Motion generated energy can include the wind, the tides and even bacteria.

Whether the energy and capital cost required to harness this motion is modest enough to make these efforts worthwhile is the critical question.

See EROEI

https://www.washingtonpost.com/local/trafficandcommuting/this-dupont-circle-sidewalk-turns-footsteps-into-power/2016/11/30/c69263f8-b020-11e6-8616-52b15787add0_story.html?utm_term=.ff5125165335

Natural Agricapital

Natural agricultural capital includes the condition of soils, the integrity of agricultural lands and landscapes and the knowledge, wisdom and productivity of farmers. (Yes, wisdom may be hard to quantify, but we know it when we see it.)

Natural agricapital can be declining while productivity is increasing if the productivity increases are coming from depleting such components of natural agricapital as soils, water tables, insect populations and traditional farming knowledge. In the UK, for example, a recent study estimated that the country only has 100 harvests left before its soils are completely depleted.

Enhancing natural agricapital is essential in order to increase agricultural resilience while boosting **carbon drawdown** from our soils.

http://www.independent.co.uk/news/uk/home-news/britain-facing-agricultural-crisis-as-scientists-warn-there-are-only-100-harvests-left-in-our-farm-9806353.html

Natural Carbon Lock-in See Carbon Lock-in

Natural Energy Reserves

The sum of the potential energy generated by wind, solar, and geothermal power. Calculating these reserves depends on judgments about efficiency, land area and cost.

Newables or Newable Energy

Renewal power sources are not renewable. Their fuel — wind or sun — is renewable. But the required solar panels, mirrors and wind turbines demand steel, concrete, power lines, rare elements, bulldozers, and other heavy equipment. As does the batteries or other devices required to address the **intermittency** problem. All this requires lots of energy to fabricate, most of which will come from fossil fuels for many years.

So perhaps it's more accurate to say that we need a transition to newables or newable energy — a mix of renewable fuel and not so renewably made materials.

See Renewables

Ninety-nine Percent Climate Doctrine

The author Ron Susskind wrote in *The One Percent Doctrine* that Vice President Dick Cheney argued that if there was even a 1% chance that Al Qaeda could get possession of a nuclear bomb, then that risk of possession should be treated as a certainty with appropriate action to follow.

Contrast this doctrine with climate chaos. Mr. Cheney and many others refuse to support acting vigorously (or at all in most cases) to mitigate the threat of climate chaos, even though 99% of climate scientists believe that climate chaos is real, is caused by humans and is already doing widespread damage to life on earth.

Perhaps, this difference in doctrine has at least something to do with Mr. Cheney's long time ties with the fossil fuel industry. Or is that unfair speculation?

NoHi

Plant and animal climate migration to escape the dangers of a changing climate tends to be either to the **No**rth, (except in the Southern Hemisphere) to **Hi**gher ground or to both. The goal is to reach climatic conditions like those found in their original ecosystems. Humans can help facilitate these survival migrations, and in increasing numbers humans will be migrating as well, in many cases in the same direction.

See Biotic Refugees

No Ice (at all) Age

When the last ice age ended 11,700 years ago, earth ice retreated but did not disappear. Sometime in the next several centuries however earth's ice may well completely disappear due to global heating, ushering in a no ice (at all) age.

NoRIMBY (No Renewals in My Back Yard).

A movement to stop wind turbines and solar panels in or near one's backyard.

The backlash to renewables is due to aesthetic reasons as well as concerns about bird kills. There's a slight problem with the bird kill argument however. Cats kill about 10,000 as many birds as turbines, and buildings and cell towers also kill many times more birds than turbines do.

http://www.treehugger.com/renewable-energy/north-america-wind-turbines-kill-around-300000-birds-annually-house-cats-around-3000000000.html

Nuclear Summer

An alternative term for global warming.

Based on the estimated 400,000 Hiroshima atomic bombs that climate change adds in extra energy to the earth *every day*, we could fairly describe global warming as a kind of nuclear summer, getting hotter and hotter by a factor of many nuclear explosions daily.

In contrast to nuclear summer, the term 'nuclear winter', popularized in the 1980s, describes a condition of cooler darker weather resulting from nuclear war. The cooling results from the massive amounts of soot sent into the atmosphere from nuclear explosions, not unlike the phenomenon of **global dimming**.

https://thinkprogress.org/earths-rate-of-global-warming-is-400-000-hiroshima-bombs-a-day-44689384fef9#.sc2g6pp1n

Nudgeatarian See Carbon Nudge

Ocean Hawks

Those who believe that aggressive action to reduce the impact of harmful forces on our oceans is essential. As the extraordinary decline in oceanic health continues unabated, the number and political visibility of ocean hawks will markedly increase.

The afflictions facing oceans are multiplying. Among the most alarming are a 30% increase in acidification, dramatic increases in coral reef bleaching, a 40% loss of phytoplankton since 1950, a reduction in oxygen levels, and an estimated 50% decrease in marine populations just since 1970.

The consequences for climate action are vastly complicated by these indicators of oceanic decline. Phytoplankton may not be able to absorb more CO_2, coral reef collapse could jeopardize as much as 25% of the marine life in the ocean exacerbating world hunger, and acidification may accelerate atmospheric warming by reducing sulfur concentrations, which in turn may diminish sulfur's cooling effect on the atmosphere.

Among the major actions recommended by ocean hawks are the

creation and expansion of marine reserves (the world's largest marine reserve was created in October 2016 in Antarctica), the banning of pollution and waste disposal, the cleanup of oceans and increased research to better understand ocean ecology. And most critically of course is to cut carbon emissions to close to zero as soon as possible.

Perhaps a black box should be placed around this entry given the grave status of the world's blue waters. Our oceans plead for our help. We must all become ocean hawks.

See Coral Reef Bleaching, Marine Heat Waves, Ocean Acidification, Reef Races

http://assets.wwf.org.uk/downloads/living_blue_planet_report_2015.pdf?_ga=1.259860873.2024073479.1442408269

Octopus Alert!

You may have to fight off an octopus in a parking garage, at least during a king tide in Miami. Given the recognized intelligence, playfulness and sense of direction they are known to possess, it may well be that the octopus truly desired to occupy a parking space. Perhaps next time you see one it will share the secret as to why it's so attracted to the cozy confines inhabited by a school of motor cars.

No word yet on what rate the octopus was charged, how many spaces it used, whether it left on its own eight feet or where it stored the entry ticket.

See Global Weirding

http://www.livescience.com/56952-octopus-stranded-due-to-supermoon-tides.html
http://www.wsj.com/articles/our-noble-cousin-the-octopus-1480714083

Pagency

A portmanteau of 'patience' and 'urgency', two qualities needed in varying quantities over time to navigate the climate crisis. Urgency as CO_2 emissions must be reduced by 80% or more in a generation to avoid a doomsday calamity, and patience as the positive effects of CO_2 reductions will take years to become evident, during which time societies must remain disciplined, focused and ordered.

The writer, activist and Buddhist scholar Joanna Macy suggests a practice she calls "deep time work," "where each of us expands the meaningfulness and impact of our lives beyond the span from our birth to our death in this lifetime, and realize that we're here at this critical juncture of our history where we could blow it all. And we're acting for and with both the past generations and the future generations."

Now that is a practice that cultivates pagency.

See Bifocal Behavior

https://www.garrisoninstitute.org/blog/finding-courage-create-new-culture/

Pascal's Climate Wager

The philosopher Blaise Pascal formulated his famous wager in the seventeenth century. If God does not exist and I believe in him, I do not lose anything. If God does exist and I do not believe in him, I go to hell for all eternity.

If climate change does not exist and I believe in it (and act on that belief), I have helped create a better world. If climate change does exist and I do not believe in it, then I have contributed to all of life essentially going to hell for all eternity.

Patient Climate Capital

Patient capital is investment for the long term, where returns are not seen for some years. Trillions of dollars of patient climate capital must be found to finance the **one generation challenge** necessary to transform our energy and other systems to a **zero-carbon** economy over a 20-year period.

Providing powerful tax, regulatory and other incentives for patient climate capital will be required to meet this challenge.

See Bifocal Behavior, Pagency

Peak Oil Debt

In 2016, the four largest oil companies had their highest collective debt ever, a function of low oil prices and the much higher cost of oil exploration. These revenue pressures will only be magnified by the **Keep It in the Ground** campaign, making it likely that major oil company debt will continue at or near record levels.

Whether peak oil debt will encourage major oil companies to downsize, refocus some of their business on renewables or simply muddle through in the hope that good times will return is a key question whose answer will affect our ability to fight climate chaos.

See DETS Crisis

http://www.zerohedge.com/news/2016-08-24/oil-debt-soars-cover-capex-investors-assert-dividends-are-unsustainable

---

Permadeath

Permafrost in the Arctic regions is beginning to thaw, leading to a risk of massive releases of CO_2 and Methane. Temperatures could increase by over 1° when permafrost is completely thawed, as it is believed to contain more than twice the CO_2 than the atmosphere, in addition to considerable methane.

Permadeath could also trigger *hyperfeedbacks* leading to greater climate chaos.

See Arctic Amplification, Climate Positive Feedback, Drunken Trees, Methane Crater, Runaway Climate Change, Tipping Points

https://www.theguardian.com/environment/climate-consensus-97-per-cent/2015/oct/13/methane-release-from-melting-permafrost-could-trigger-dangerous-global-warming

Pet Projects

People often value their pets more than they value the lives of other people and perhaps even their own lives. Some 90% of pet owners consider pets to be part of their family. One survey showed that 40% of married women get more emotional support from their pets than they receive from their husband or kids.

Projects that target such very real threats to pets from climate change as scorching heat, severe drought and flooding, disease outbreaks and food shortages, are likely to motivate owners to support climate action.

Placebo Plan

The common practice of setting modest carbon reduction targets for the near future — generally 10 to 15 years — and much more ambitious targets for the distant future, typically 30 or more years out to 2050 or so.

This backloading of deep and necessary climate reductions, while politically and economically convenient for the organization or politician, may be unhelpful for the climate. The longer we wait to make draconian reductions, the more CO_2 builds up and exacerbates climate chaos.

Another dodgy or placebo maneuver is the use of a high baseline year (baseline bias) for carbon emissions so that the targeted reductions will be much less than would be required if a low carbon emissions baseline year were chosen.

See Climate Horizons, Procrastination Penalty

Planetarian

One who cares strongly about and identifies with the welfare of the planet. In most contexts 'planetarian' can be used as a substitute for environmentalist, given the negative connotation that environmentalist has for many people. Besides, caring for and protecting the planet is the job of all of us, not just those who label themselves or are labeled by others as environmentalists.

Planetary Digital Afterlife

See Civilizational Preservation Movement

Planetary Panic Points

As the reality of irreversible climate chaos becomes evident, a series of behavioral tipping points or panic points are likely to occur.

An example would be the panicked attempt to sell property near the sea once the realization that sea-level rise is occurring now and can't be stopped.

Another planetary panic point may result from increasingly severe water shortages in countries dependent on water from the third pole. Water hoarding and eventually mass migrations will ensue.

As with financial panics, on the surface everything seems fine until a collective consciousness emerges that something critical is about to collapse. At that point it's usually every person for themselves.

The US may be close to a panic point with respect to east coast seashore property. Property values are beginning to suffer as people experience an individual or collective **oh shit moment**.

The November 24, 2016 New York Times article below might accelerate the collapse in coastal values, as it may be the most prominent major story about coastal sea rise and its impact on property values and consumer psychology.

See Climate Canary, Tipping Point

http://www.nytimes.com/2016/11/24/science/global-warming-coastal-real-estate.html?&target=comments&hp&action=click&pgtype=Homepage&modref=HPCommentsRefer&clickSource=story-heading&module=second-column-region®ion=top-news&WT.nav=top-news&mtrref=undefined

---

Planetary Prosthetic

Products, services, processes or techniques that can enhance the earth's health.

Plan-it Earth

This spelling of planet earth reminds us of the importance of consciously planning our future to survive climate chaos.

Plant Purgatory

Plant viability may be the critical factor in determining human survival in a **post normal climate**. We can migrate from toxic heat and sea level rise; we can apply our knowledge of public health and medicine to minimize climate related disease; we can redesign our settlements and structures to conserve water and protect against high winds. But we can't find substitutes for food.

Higher temperatures, greater CO_2 concentrations in the air, land and sea, changing rainfall patterns, emerging crop diseases and decreases in soil fertility all are likely to affect the ability of our staple crops — rice, corn, wheat and soy, as well as other crops, vegetables and animals — to grow and thrive. This in turn may lead to mass starvation, panic, and social chaos. Our current national obsession with the evils of junk food may soon be overshadowed by our concern with getting enough of any food.

http://www.takepart.com/feature/2016/11/28/climate-resilient-plant-breeding

Polar Bear Propaganda

Polar bears are perhaps the species most identified with the adverse consequences of climate chaos as a result of any number of public relations campaigns. However, questions can be raised about the effectiveness of defining polar bears as a climate chaos mascot.

According to George Marshall in *Don't Even Think About It: Why Our Brains Are Wired To ignore Climate Change*, polar bears, by living in such remote and exotic locations, may not be an effective symbol for a problem that is made so difficult to tackle in part by its remoteness.

According to Marshall, the charity Christian Aid chided polar bear symbolism in a poster that showed, on one side, Africans on cracked earth, and polar bears on the other side, with the caption "Climate Change threatens more than just polar bears and ice caps."

A retired Army motor transport specialist who experienced the dangers of transporting fuel in fuel convoys in war zones was featured on an episode of the television series Years of Living Dangerously, saying that clean energy "is not about polar bears, it's about fuel convoys and national security."

I don't imagine that you'll be surprised to hear that humans masquerading as polar bears don't do well as motivation for climate action either, per a 2016 study done by Climate Outreach.

See Eat Your Carbon Spinach

Pogo Planetary Paradigm "We are the enemy and he is us."

Perhaps the most apt aphorism of the climate chaos era.

Human beings are responsible for essentially 100% of climate chaos and since the 1980s we have known that it is we who are responsible.

See Central Climate Conundrum.

Post Normal Climate

A post normal climate is a climate that is unpredictable, novel, often extreme and ever changing, almost always for the worse.

While 'new normal' is often used to describe the emerging climate, post normal conveys a degree of change and unfamiliarity beyond even that of 'new normal'.

See Never Normal Coalition, Predictable Unpredictability

Pre-mourning

The feeling of anticipated loss. A soon to be common climate chaos induced feeling.

See Ecocholia, Oh Shit Moment, Solastalgia

Prolonged Pollen Penalty

In parts of North America and elsewhere, the hay fever season has lengthened by up to a month largely due to changes in temperature and precipitation induced by climate chaos.

https://www.scientificamerican.com/article/us-price-tag-for-allergies-will-rise-because-of-climate-change/

Punctuated Disequilibrium

Short spurts of particularly intense climatic chaos may occur in the midst of few periods of real stability.

The theory that evolution is marked by long periods of relative stability interrupted (punctuated) by short periods of rapid species change is called punctuated equilibrium. Applying this concept to climate chaos suggests punctuated *disequilibrium* may be a more accurate way to describe climate change over time.

See Post Normal Climate

Readaptation

It may be more accurate to speak of (re)adaptation than **adaptation** in a **post normal climate.** Adaptation actions, be they flood barriers, home designs to protect against storms, or educational programs to deal with climate stress will rarely be able to achieve their purposes for long as the climate will change frequently in highly unpredictable and extreme ways. Thus, the need to periodically revisit and revise adaptation measures to insure their continued effectiveness.

See Global Weirding, Predictable Unpredictability, Punctuated Disequilibrium.

Resettlement reserves

A critical component of effective disaster planning is the early identification of a range of communities that can serve as reasonably safe locations for resettlement.

Given that millions, perhaps tens of millions will need to be resettled from Southern Florida, New York, Boston, New Orleans and other coastal areas this century serious planning for the creation of resettlement reserves must begin before too long.

Possible resettlement reserves may be military bases, summer camps or monasteries, abandoned airports and other large tracts of land. Many communities in more climate secure areas should be given the opportunity to compete to house resettlement reserves, and be financially supported by the Federal government.

See First Flee/ers, Life Zones

https://www.livescience.com/54042-climate-change-could-force-coastal-retreat.html

Road Closures Ahead

Maintaining much less expanding our road network becomes increasingly counterproductive in an era of climate chaos. Closing roads will reduce the maintenance and capital burden on governments. It may reduce the number of vehicle miles traveled. Tearing up asphalt will increase **albedo**.

While most roads will have to be maintained because of the need for local access many can be put on a 'road diet' to reduce surface area, and increase access for bicycles and pedestrians.

Of course, it is easy to argue for road closures. However, a recent study projected that some some 25,000 kilometers of new road may be built by 2050.

The ability to meet our climate goals will be much more difficult to meet if this volume of roads is allowed to be built.

See Billion Car Campaign, Climate Defense Interstate Mobility System, Transport Triage

http://www.nature.com/nature/journal/v513/n7517/abs/nature13717.html

---

Sacrifice Planet

The destruction of life on earth resulting from climate change and other environmental insults is rapidly leading to the entire earth becoming a sacrifice planet.

Based on the concept of sacrifice zones, or areas generally inhabited by lower income people that are so degraded by industrial or other activity as to make them virtually or actually uninhabitable.

Seasonal Cities

Cities that may be inhabited only at certain times of the year because of excessive and dangerous heat and humidity levels or because of the likelihood of seasonal severe flooding.

Many cities in the Middle East and South Asia are beginning to experience extreme summer time temperatures that make daily outdoor activities close to impossible. How much higher will these temperatures rise before these cities will need to shut down for periods of time, particularly if energy shortages or fossil fuel restrictions limit the use of air conditioning.

What might be called receiving seasonal cities are those communities in the northern latitudes or high above sea level that may experience much greater seasonal migrations from people looking to escape heat stressed cities.

Sequential Alarmism

In our **post normal climate,** the planet is being subjected to rain bombs, droughts, floods and extreme heat with increasingly greater unpredictability and frequency. This combination of frequency, intensity and most critically unpredictability will lead to what might be called sequential alarmism as each climate event unfolds.

See Never Normal Coalition, Post Normal Climate, Predictable Unpredictability

Silent Eemergency

The world is entering an **eemergency** situation characterized by a societal silence that we are in an eemergency situation.

See Media Omertà

Slow Movement Movement

This potentially far reaching movement applies "slow" principles in such areas as eating, mobility, parenting, money and almost every other area of life.

Slow movements decrease consumption of fuel and lead to fewer and less severe accidents. More broadly an enhanced sense of well-being and resilience results from a slow mindful presence.

As the writer Paul Hawken has said:

> We are speeding up our lives and working harder in a futile attempt to buy the time to slow down and enjoy it.

See Degrowth, Plenitude

Smart Degrowth

An approach to achieving a comprehensive quantitative downsizing of society in thoughtful, humane and comprehensive data driven ways. Derived from smart growth.

See Degrowth

Southern Sink

The waters of the Southern Ocean encircling Antarctica play a central role in regulating the earth's climate through their ability to act as a major carbon sink. Scientists estimate that these waters may have absorbed as much as 15% of the CO_2 emissions humanity has poured into the atmosphere since the industrial revolution.

The strength of the sink has varied greatly in recent decades, even becoming a source of CO_2 for a decade or so starting around 1997.

Intense research efforts are now focused on uncovering the many gaps in our knowledge of how the Southern Ocean affects the world's climate.

See Sources and Sinks

http://www.nature.com/news/how-much-longer-can-antarctica-s-hostile-ocean-delay-global-warming-1.20978

Supermom (or Dad) Paradox

It is documented that people can lift a car that weighs thousands of pounds in an emergency to save their child (or someone else's).

Yet millions and millions of parents generate high levels of carbon emissions through their consumption patterns apparently never thinking twice that this consumption is worsening climate change. Which in turn is dooming their kids (as well as yours and mine) to a bleak existence in a **post normal climate**.

See Arctic Ice Sensitivity or Claim Your Arctic Ice While It Lasts, Carbon Consumption Conspiracy

https://www.psychologytoday.com/blog/extreme-fear/201011/yes-you-really-can-lift-car-trapped-child

Syntheticism

Syntheticism is a term for the Ecomodernism environmental policy movement formulated by the Breakthrough Institute in Oakland, California.

This idea is consistent with technologies and practices such as nuclear power, urban densification, genetic engineering, intensive farming practices and liberal use of GMO's. Syntheticism sees these as not only desirable but essential to minimize environmental degradation and climate chaos.

Syntheticism can be contrasted with naturalism, which is the extensive use of land for food, energy and raw material production as well as for human settlement, utilizing natural materials and practices wherever feasible.

These two philosophies present contrasting alternatives for organizing a society. It may well be that a pragmatic evidence based synthesis of naturalism and syntheticism is the best approach to minimize climate impacts, spare nature and address global poverty.

See Bright, Dark and Light Green Environmentalism

http://www.ecomodernism.org

Tack Mastery

The art of bringing a system back to true course by skillful navigation of the elements. The uniqueness of tack mastery is its recognition that the fastest and sometimes only way forward is by navigating on carefully calculated angular pathways rather than straight ahead.

Given the multiple complex and often counterintuitive feedback loops that characterize climate chaos, skillful tack mastery may be an essential perspective and skill. The term is derived from the sailing term 'tacking' which means angling into the wind to make forward progress.

See Carbon Maze

Terragasm

A cataclysmic upheaval occurring on, under or above the earth, particularly if caused by or made more likely by climate chaos.

See Climatastrophe, Gigatastrophe

Terramortis

Death of the planet.

See Biomortis

Time Ton Dilemma

Every year we delay reducing CO_2 emissions into the atmosphere we use up millions of tons of the remaining **carbon budget** and reduce our already slim chance of stabilizing temperatures at or below 3.6°. The dilemma is that each unit of delay requires a correspondingly greater percentage of carbon emission cuts to achieve the same temperature goal.

See Friedman Climate Unit, Procrastination Penalty

Transplant Nations

Those nations located in the Pacific or Indian Oceans, that because of low lying lands and sea level rise may be unable to retain the land mass to survive.

These nations may attempt to retain their national cohesion, institutions and culture by transplanting themselves in as intact a state as possible to another land, subject to agreement with the receiving country.

Questions of national sovereignty of the transplanted nation, international recognition, degree of autonomy, currency, compensation and economic and cultural integration will need to be sorted out without the benefit of much direct precedent.

Physical, economic, political and legal planning for this eventuality may need to begin in the near future.

Kiribati, the Marshall Islands, the Maldives and Tuvalu appear to be the island nations most at risk and thus most likely to consider becoming transplant nations.

2 + 2 =\= 4 (Celsius that is)

Projections done for the IPCC and others show that a 6.4° increase in temperatures this century is plausible under business as usual scenarios. Since the impacts from climate chaos do not change in a linear fashion 6.4° represents a devastating and likely fatal situation for the human project on this planet. To state it simply, as German climate scientist John Schellhuber does, "The difference between two and four degrees is human civilization."

2° C Dogma

The generally recognized upper limit for a safe projected global temperature increase since the start of the industrial era is 3.6°. (2° Celsius)

This target is more dogma than science. There is no rigorous scientific justification for this or any other similar level of warming being the limit beyond which temperatures would make a livable world impossible to maintain. (In fact, the 3.6° threshold was proposed by a Yale economist - William Nordhaus - in the 1970s.) Yet 3.6° remains a kind of worldwide mantra invoked to reassure people that the planet still has some way to go before it crosses this "critical" threshold.

Many scientists believe the world has already locked in a 3.6° rise and even those who don't believe we've locked in the limit do believe we have only a very few years to drastically cut emissions to stay on the 'safe' side of 3.6º. Others argue that havoc is occurring now when we're below 2.5° so how can we be confident that we can live with a 3.6° world.

Some analyses conclude that we have much less than a 50% chance of staying under this 3.6° limit. How many of the Paris Climate Agreement signatories would board an airplane if it were known that it had less than a 50% chance of landing safely? Given the almost unthinkable consequences of a greater than 3.6° world how many of the rest of us would think the agreement resulted in an "acceptable" risk.

Yet the leaders of **COP21** cheered as the climate agreement was announced. We are a peculiar species.

See Carbon Budget, Tipping Point

http://www.pbs.org/newshour/bb/why-2-degrees-celsius-is-climate-changes-magic-number/

Universal Repricing

Widespread climate chaos is likely to affect the supply and demand of virtually every product and service on earth. As the climate changes, so will prices and currencies.

Stable currencies and prices are a critical component of the social order. Their absence will threaten social cohesion and add yet another challenge to the many we face in adapting to climate chaos.

Waterhoods

Neighborhoods regularly inundated with floods.

White Only Laws

A proposal to have a 'white only' rule or law would require that surfaces such as sidewalks, streets, rooftops and building façades be as white as possible (except when other surfaces such as green foliage may provide equal or greater reflectivity and other benefits).

Perhaps the Henry Ford mantra of being able to purchase any car as long as it is black will be modified to being able to purchase any car as long as it is white.

While the climate cooling related benefits of white surfaces in cities have been identified not all studies concur. One study concluded that light reflecting off white surfaces will interact with atmospheric particles and possibly increase heat absorption rather than reducing it. And in parts of the country or the world requiring heating in the winter, white roofs are likely to increase heating requirements and thus increase carbon emissions.

Given the uncertainty about these impacts we may be seeing something akin to a summer camp phenomenon — color wars — break out, with white, green and black (the de facto color of most roof and road surfaces) dueling for primacy in our cities.

See Albedo, Carbon Maze, Heat Islands

http://www.climatecentral.org/blogs/white-roofs-may-increase-global-warming

---

Zombie Bacteria

Bacteria in suspended animation due to frigid temperatures can come back to life as temperatures increase.

A chilling (no pun...) example of this was the outbreak of anthrax in Siberia in 2016. The permafrost thawed, which led to the thawing of reindeer carcasses which had been colonized by anthrax bacteria some 75 years ago. Both reindeer herds and humans were then infected by anthrax with some 1500 reindeer and 13 humans sickened and one boy killed.

See Permadeath

http://www.bbc.com/news/world-europe-36951542

Climate Proposals

BDR Movement (Boycott Divest and Revest)

A proposal to create a movement that boycotts fossil fuel companies, divests from fossil fuel holdings and reinvests (revests) in clean energy and other activities that provide carbon benefits.

This effort to stay within the world's **carbon budget** is a broader and more comprehensive strategy than the current fossil fuel divestment movement, which currently includes institutions with collective assets of some $5 trillion.

https://insideclimatenews.org/news/13122016/fossil-fuel-divestment-movement-climate-change

Billion Car Challenge

A proposal to create a worldwide campaign to sharply downsize the planetary automobile fleet.

It is estimated that there are over 1.2 billion cars in use on the planet. Of that number, less than 3% are electrically propelled. To reduce carbon emissions by well over 80% in one generation, the world will have to transform its fleet in profound ways.

Some combination of substantially reducing the number of cars, (rather than seeing the fleet go to two billion as countries such as China continue to become more affluent) switching cars to electric and other 'alternative' power, shrinking them to reduce carbon emissions from their manufacture and operation, and radically improving gas mileage of the remaining vehicles will be necessary. Self-driving cars and car sharing services will be a key part of this transformation.

Car companies, the **carbon combustion complex** and national governments will need to spearhead this massive effort.

See Assisted Movement Systems, Transport Triage

Building Biopsy

A proposal for the development of analytical tools to examine the components of a building to determine ways to reduce carbon emissions or embodied carbon. A building biopsy can help guide the design, construction and operation of **zero carbon buildings.**

C Corp (Climate Corp)

A proposal for a new type of corporate charter that would empower corporations to undertake vigorous action to address climate chaos. C Corp status would allow corporations to place climate actions above all other interests.

C Corps requirements could include mandatory climate audits and commitments to attain zero or negative carbon emissions within a certain time frame. Tax and regulatory incentives could be developed to encourage companies to organize or reorganize as C Corps.

The recently established B Corp, the closest model to the C Corp, authorizes corporations to consider all stakeholders, not just stockholders. (Regular corporations are often known as C Corps.)

Carbon Canary Project

A proposal to identify and publicize early warning indictors of climate damage.

These could include behavioral changes, death or morbidity of certain species, economic or behavioral anomalies experienced by people or communities, or almost any other indicator that provides warning to the world community of the need to act on climate chaos.

An agreed upon suite of indicators should be regularly and widely publicized so that all the world would know the status of our fight to rein in climate chaos.

See Planetary Panic Points

Carbon Cap Commissions (CCC)

The world must set firm caps on carbon emissions to avoid severe climate chaos.

One proposed mechanism to achieve this is the establishment of regional, national and international Carbon Cap Commissions. These government established agencies would have the responsibility to administer carbon cap limits. The caps could be designed as cap and trade programs, tradable emissions quotients or other mechanisms.

While setting the level of the caps and enforcement penalties would likely be a legislative responsibility, CCC's would have a critical role in guaranteeing that the public perceives these caps as transparent, fair and effective.

Carbon Concentration Clock

A proposed public display or clock showing the concentration of carbon in the atmosphere. The clock would be analogous to the national debt clock that was in Times Square in New York City.

Perhaps some New Year's Eve revelers will take a moment, as they view the carbon clock, to reflect on the number of years left where conditions will be conducive enough to even be outside on New Year's Eve.

See Keeling Curve

Carbon Crimes Against Humanity

A proposal to adopt laws that define carbon crimes against humanity.

Active climate chaos denial or promotion of fossil fuel burning may qualify the perpetrator as committing carbon crimes against humanity.

See Carbon Complicity Index, Carbon War Criminal

Carbon Dye See Carbon Signature

Carbon Exclusion Zones

A proposal for areas where people and carbon related economic activity is prohibited.

Exclusion zones currently include nuclear hazard areas near Chernobyl or around active volcanoes. As the **Keep It in the Ground Campaign** gains momentum and the extremely tight **carbon budget** is taken seriously thousands of carbon exclusion zones could be established around potential fracking sites, oil wells and refineries, major pipelines and coal mines.

Natural areas with large carbon concentrations such as peat bogs and parts of the Amazon could also be designated as carbon exclusion zones.

See Carbon Lockbox, Carbon Retirement, Intentional Grounding

Carbon Lockbox

A proposal for a secure mechanism to ensure that sequestrated carbon does not escape into the atmosphere. The lockbox is based on the expression used by Vice President Al Gore to describe a way to prevent Social Security funds from being used for non-Social Security purposes.

The lockbox may be an actual physical arrangement or a set of security measures to ensure that fossil fuels are not burned, a set of criminal or civil penalties, and/or financial or other incentives.

How to prevent other sources of carbon from entering the atmosphere such as the burning of trees, carbon intensive agricultural production or methane releases from cows and other ruminants is a more challenging problem not easily solved by the carbon lockbox.

See Carbon Exclusion Zones, Carbon Retirement, Carbon Signature, Intentional Grounding

Carbonoscopy

A proposal for a device or mechanism that can examine machinery or systems in order to help minimize their carbon emissions. While the process may not be as physically unpleasant as a colonoscopy, the results can be equally significant in discovering 'carbon cancers' that need to be resected.

Carbon Reserve Bank

A proposal to create a public or public/private bank that could undertake climate financing, as well as the development and administration of carbon taxes, credits or caps. The Carbon Reserve Bank is modeled on the Federal Reserve Bank.

One hopes that the markets will look as closely at the bank's carbon regulations as they look at the Federal Reserve's interest rates pronouncements.

See Carbon Cap Commissions

Carbon Signature

A proposed process that enables one to identify the source and other characteristics of particular carbon emissions, perhaps through use of a carbon dye.

Carbon STOP **S**ubsidies **T**echnology **O**pinion **P**ricing

A proposed campaign to curtail carbon emissions.

The four legs of the campaign are stopping subsidies for carbon burning, accelerating investments in clean energy technology, supporting media and educational efforts to mobilize public opinion towards climate action, and putting a price on carbon by a tax, fee, or cap and trade system.

CDARPA The Climate Defense Advanced Research Project Agency.

A proposal to create a climate research organization staffed by elite scientists to develop and advance technologies and processes to mitigate and adapt to climate chaos.

CDARPA is modeled after the Defense Advanced Research Project Agency (DARPA), which has a storied reputation for developing such transformative technologies as the internet, GPS and stealth aircraft.

Placing such an agency within the Defense Department will allow the military to play a key role in dealing with the causes and effects of climate chaos. Giving the military a key role has advantages. It is already aware of climate threats and is acting to address them; it has a high level of public acceptance as well as a long and impressive track record in successfully conducting research and translating it into workable technologies.

Folding the U.S. Advanced Research Projects Agency–Energy into CDARPA would bring most climate related research under this one umbrella.

Clibernetics

A proposal to establish a new discipline focused on climate change communication.

Climadelic Therapy

A proposal to legalize psychedelics for use in helping people to cope with the emotional dimensions of climate chaos.

Emerging research is showing the ability of psychedelic substances such as LSD and psilocybin to help people overcome post-traumatic stress disorder, depression, the anxiety one feels facing cancer and a number of other mental health conditions.

These effects go well beyond symptom relief, important as that is. Most participants in these studies maintain the benefits of psychedelic substances long after they stop using them. Often these benefits occur after a single dose. People experience profound changes in their world-view, becoming more optimistic, more grateful and appreciative of life, more spiritual, and more accepting of what life has in store for them.

The potential for climadelic therapy to help people accept the reality of climate change with less panic, despair and existential dread and with more equanimity and confidence is almost unlimited.

Yet climadelic therapy can go beyond helping people maintain equanimity in a **post normal climate.**

Love and connection rather than material possessions is more likely to become the focal point of one's life after climadelic therapy. And that's good for the person fortunate enough to experience that and for a society that needs to shift away from goods to good.

Research should be accelerated with legalization as soon as possible.

See Climate Fatalism, Collective Empathy, Oh Shit Moment

http://www.newyorker.com/magazine/2015/02/09/trip-treatment

Climademic

An academic whose primary expertise is climate chaos.

See Carbon Bariatrician, Clibernetics

Climataceuticals

Products designed to assist in preventing or easing symptoms or conditions resulting from the effects of climate chaos.

Among these might be products that cool the skin, protect eyes against severe glare, and increase immunity to insect or water transmitted disease.

One example of a climataceutical was created in 2016 at Stanford University. Scientists developed a fabric that significantly minimizes the heating of the skin from high temperatures. This "personalized cooling" may also have applications in tents, buildings and vehicles.

See Natick Climate Center

http://www.latimes.com/science/sciencenow/la-sci-sn-cool-shirt-20160901-snap-story.html

Climate Adjusted Rate of Return (CARR)

A proposal to create a standardized methodology to incorporate climate related investments into rate of return calculations.

Whole cost accounting approaches would require this kind of calculation in order to ensure that potential investments fully value the costs and benefits of climate change related impacts.

This tool retains the widely-accepted approach of evaluating investments based on their rate of return on capital while requiring that climate considerations be factored in through a rigorous and standardized process.

Climate Cognoscopy

A proposed probe conducted through scans, tests, interviews, and other means into the minds of people to determine the nature of their climate chaos beliefs, skills, and state of mind.

Based on the term *cognoscopy* formulated by neurologist Dr. Dale Bredesen, referring to a complete cognitive assessment, analogous to a colonoscopy for the gastrointestinal tract.

See Carbonoscopy

Climate Complete Communities

A proposed program that would support and certify existing and planned communities that incorporate a full range of climate mitigation and adaptation features in their design and management.

These would include zero carbon buildings and transport systems, onsite renewable sources of energy, systems to protect against climate hazards such as flooding and excess heat, and robust citizen participation efforts.

Climate Connectomics

A proposed field of study that would examine the connections between climate theory, data and action to determine the most effective way to fight climate chaos. Derived from the field of Connectomics, which diagrams the connections within the nervous system and brain.

See Carbonoscopy, Clibernetics, Climate Cognoscopy, Integrated Climate Planning and Management

Climate Counseling Corps

A proposed nonprofit or governmental entity whose mission is to provide mental health services and emotional support for those who are affected physically and emotionally by climate chaos.

The CCC could become the lead organization in providing training and support to millions of people to assist them in coping with climate chaos.

See Climadelic Therapy

Climate Equity Guarantee Agreements (CEGA)

Proposed agreements that compensate a company for losses it may suffer as a result of rising sea levels and other consequences of climate chaos, in return for a commitment to climate adaptation measures, relocation commitments, employment guarantees or other appropriate actions benefiting the community.

Imagine the human and financial chaos when all or portions of a city or town may need to be evacuated or relocated. CEGA's can be a mechanism to encourage stability and continuity of operations.

See Climate Insurance

Climate Extension Agents

Climate university trained and supported agents would provide a broad range of technical support to business and households, including energy efficiency and renewables planning and resilience training.

Climate extension agents are based on the very successful agricultural extension agent model. They would work closely with the **Climate Counseling Corps**.

See International Climate Corps

Climate Hedge Funds

A proposal for the creation of funds modeled after conventional hedge funds, but focused on investments that seek to gain from climate chaos linked economic disruptions or climate supportive income opportunities. The proceeds from these funds could be taxed or otherwise directed towards climate mitigation activities.

Climate Justice System

A proposal to define a set of climate infractions and enforcement protocols and penalties. Special attention would be given to **climate truth and reconciliation** efforts, **carbon piracy**, and policing protocols for various kinds of climate related infractions.

See Carbon Police

Climatekeeper Alliance

A proposed nationwide and ultimately worldwide network of citizen groups focused on protecting the climate in each community or region.

Modeled after the Waterkeepers Alliance and various Riverkeeper organizations each local Climatekeeper group would define its own priorities and create an action plan to tackle climate change mitigation and adaptation.

Climate Pulitzers

A proposal to create a Pulitzer or Pulitzer like prize to be awarded to those who write or publish the most impactful articles on climate chaos.

Establishing high profile awards encourages the often 'hear no evil see no evil' media to pay attention to the most important story of all time.

See Kardashian Climate Ratio, Media Omertà.

Climate Richter Scale

A proposal to create and utilize a numerical scale to rank climactic disturbances. Based on the widely-accepted Richter scale that ranks earthquake intensity.

Climate Tithing

A proposed means of fundraising where people or institutions donate 10% of their incomes to organizations or causes fighting climate chaos.

Climate Trustee

A proposal to adopt laws mandating the creation of Climate Trustees who would represent the interests of the young and the unborn in fighting climate change. These trusteeships could be created by all levels of government.

Powers of the Climate Trustee could include the ability to bring lawsuits, testify at administrative and legislative proceedings, issue public reports and proclamations, and have official status as interveners when climate issues are being deliberated.

The Trustee would also participate in cabinet and other policy and regulatory meetings that are otherwise closed to the public. Allowing the Trustee to bring young people to selected meetings could help further insure that their interests are fully considered.

See Children's Climate Crusade

Climate Truth and Reconciliation Commission

A proposal to create a public body to investigate individuals and organizations that lied, misled or otherwise delayed public or private action on climate chaos. Public hearings would allow people to testify about how they have been or expect to be harmed. Witnesses would explain what the climate deniers did and the impact of their actions.

Based on the Truth and Reconciliation Commission established in South Africa after the fall of apartheid, such an effort is intended to accomplish just what the name suggests — to bring the truth out about who helped block action on climate chaos, find ways to reconcile with the miscreants and unify the nation in its efforts to address climate chaos.

See Carbon Crimes Against Humanity, Climate Justice System, Climate Recriminations

Climate Universities

A proposal to establish a network of major climate universities throughout the world. These universities would focus exclusively on all aspects of climate chaos. A triad of teaching, research and service to the larger community, region and world would guide academic programs and other activities.

Climate universities would provide a much needed inter and multi-disciplinary focus on climate chaos. As with so many other problems in today's advanced societies the extreme compartmentalization of knowledge thwarts effective action. Climate universities would be organized to maximize collaboration and minimize compartmentalization.

Long term focused research, often done in cooperation with other climate universities, would provide the world with much-needed comprehensive knowledge of climate chaos. The service mission would provide a wide variety of educational and technical resources to communities and individuals.

Foundations, governments, corporations and private citizens would be asked to contribute to their establishment and ongoing operations.

These institutions would allow young people to engage directly in the challenges of addressing climate change, and may even provide a partial antidote to the **distraction dilemma**.

If any climate universities choose to field a football team they may benefit from not being penalized for **intentional grounding.**

Climetrics

A proposed discipline that would focus on the metrics of climate chaos. Among its goals would be the reduction of climate innumeracy.

A particularly important focus of climetrics should be developing protocols to insure the accuracy, honesty and transparency of emissions and other critical data compiled by national governments.

Whether through limited technical capacity or calculated dishonesty the potential for inaccurate and misleading data to be generated and reported is a significant concern.

Generally Accepted Climate Accounting Principles (GACAP)

A proposed modification of GAAP (Generally Accepted Accounting Principles) to incorporate **whole cost accounting, internal carbon pricing**, and other climate related data and practices.

Grand Climate Bargain

Given the standoff between the majority that understands the need for prompt action on climate chaos and the denier minority, the need for some kind of 'grand climate bargain' is becoming more urgent.

What that bargain might look like is unknown. Perhaps regulatory controls on carbon emissions will be eliminated in return for the imposition of a revenue neutral carbon tax or fee.

Much will depend on whether the denier minority is more interested in addressing an existential crisis or continuing to deny that a crisis even exists.

See Denier Triad

https://niskanencenter.org/blog/case-scott-pruitt/?utm_source=Digest&utm_campaign=cdb45d1178-EMAIL_CAMPAIGN_2017_01_24&utm_medium=email&utm_term=0_b57b3ffb8c-cdb45d1178-334782285

Integrated Climate Planning and Management (ICPM)

An approach to climate planning and action that is holistic and comprehensive. All relevant economic, social, political and behavioral factors are considered in developing strategies to minimize climate damages. A holistic approach is contrasted with conventional decision-making or what might be called climate compartmentalization.

An example would be an analysis of how fast to recommend that carbon emissions be reduced. Such an analysis would be informed not just by the potential climate consequences, but also the economic consequences (quite negative the faster emissions reductions occur), the political (more difficult to get assent for a faster reduction), behavioral (how to induce people to support or at least not oppose the stresses of steep cuts) and so forth.

For a city or a nation to develop an ICPM plan would require careful attention to data collection and analysis, a means to build capacity to implement the plan, community engagement strategies, and the identification and development of financing and funding opportunities. Finally, a continuous system of monitoring and evaluation is necessary to adjust the plan as necessary.

Ideally, interdisciplinary teams with experience in all these realms would be established to guide the ICPM efforts.

Integrated climate planning and management is an essential tool, not just for governments but for any organization, large or small that is responsible for addressing climate change.

International Climate Corps (ICC)

A proposal to create an organization like the US Peace Corps focused on providing climate prevention and adaptation assistance to countries throughout the world through the use of volunteers. The volunteers would essentially function as climate missionaries.

The ICC would become a powerful vehicle and symbol of international determination to tackle climate change, and provide a wonderful opportunity to unleash the idealism of the world's citizens, particularly its youth.

Irritability Index

As temperature, humidity and rainfall frequency and intensity increase irritability increases. Research has documented the increase in unusual and aberrant behavior and crime when temperatures increase. To give people a heads-up an 'irritability index' will be created to complement the 'it feels like' weather comfort index.

See Heat Harm

Kardashian Climate Index

Some statistics has shown that Kim and the other Kardashians usually receive more internet searches than 'climate change' does. One way therefore to measure the salience of climate change is through the creation of a Kardashian climate index to chart the relative popularity of these two phenomena.

Of course, an increase in climate change searches relative to Kardashian searches may be less about the increasing awareness of climate and more about the waning popularity of one or more of the Kardashians.

Life Zones

Those communities or regions certified as habitable enough to withstand climate chaos related shocks.

These life zones would have to demonstrate a high degree of **readaptation** planning and could become receiving areas for climate related migration.

See First Flee/ers, HiNo

Modern Climate Monetary Action (MCMA)

A proposal to issue currency (print money) to finance climate **mitigation** and **readaptation** activities.

MCMA is based on Modern Monetary Theory, a theory that argues that sovereign governments can never run out of money. In a world with massive unemployment and the urgent need for many trillions of dollars of investment in renewable energy, city and agricultural transformation and readaptation, why would we not print the money we need, given record low inflation levels?

Those who oppose this approach should be asked to articulate their plan for raising the money necessary to help insure a reasonably habitable world.

Natick Climate Center

The creation of a comprehensive research center to develop products and services that minimize the uncomfortable, dangerous and potentially lethal effects of climate chaos.

Named for and based upon the Natick, Massachusetts Military Research, Development and Engineering Command that is dedicated to researching ways to help insure that soldiers are safe, effective and comfortable.

See Carbon Tourniquet, Climataceuticals, Climate Palliation

National Balance Sheet (NBS)

A proposal for the creation of a NBS that would show the quantities and qualities of our natural and man-made assets over time.

The NBS would be an essential indicator of the impacts of climate chaos.

Natural assets include water systems and natural features such as mountains, wetlands and ecosystems. Examples of human assets include the capital stocks of our infrastructure systems, the state of housing, and the health and education levels of the population.

The NBS would complement GDP (or perhaps the GDP should

complement the NBS) as an indicator of the health of the nation. We will have achieved a modicum of ecological awareness when the national balance sheet announcement is given as much attention as the latest GDP announcement.

National Climate Defense Interstate Mobility System

A proposal to reimagine and recreate our transportation network to reflect climate chaos related priorities. A climate related mobility system should focus on **zero carbon** modalities and be sensitive to community preservation and environmental concerns.

Emerging modalities such as car sharing and automated vehicles as well as climate supportive modes of travel including bicycles, walking and local and interstate transit should be the core of this network.

This planned system should reflect the goals and strategies incorporated in the **National Climate Partnership Plan.**

See Billion Car Campaign, Road Closures Ahead, Transport Triage

National Climate Conversations

A proposal to initiate a series of ongoing conversations throughout the country to engage millions of residents in broad ranging discussions on all aspects of climate chaos — causes, consequences, remedies and citizen action.

While these conversations should be coordinated at the national level the specific design should be determined locally to reflect community values and to ensure that they will be productive and well received.

The best ideas emerging from these conversations should be incorporated into action planning throughout society, including the **NCPP** (See next entry).

National Climate Partnership Plan (NCPP)

An integrated effort to be led by the White House to define goals, objectives and actions with the intent of reducing carbon emissions, preparing the country for the impacts of climate chaos and leading a unified worldwide climate campaign.

The *President's Climate Action Plan* developed by the Obama Administration in 2013 is the closest the country has to a National Climate Plan. Unlike the proposed NCPP the *Climate Action Plan* is quite brief at 21 pages, and is absent detailed plans for such sectors as buildings, transportation, state and local government, citizen education and many more.

The absence of a bold and well publicized national plan is one of the obstacles to a full-scale mobilization of the population to tackle climate change. The NCPP can provide the blueprint for coordinated action at all levels of government and sectors of society.

A NCPP prepared by the Trump administration would, alas, likely to be little more than a roadmap to dismantling climate action.

See National Climate Conversations, One Generation Challenge

National Discount Rate

The quantification of the degree by which a nation values the future vs. the present. The higher the discount rate (i.e., 5% vs. 2%), the more the value of a future action is reduced in comparison to the equivalent action in the present. The selected discount rate is used to calculate the cost benefit ratio of major infrastructure and related projects.

An example of a discount rate for climate investments is the 1% rate that Lord Nicholas Stern incorporated into the *Stern Review on the Economics of Climate Change*, a comprehensive report for the British government released in 2006.

Lord Stern argued that our grandchildren's lives and welfare should be valued about as much as ours today. (A totally equal future valuation would have a 0% discount rate. His argument for 1% is that there's a chance that we're wasting resources on the future if society isn't in existence then — a not outlandish probability.)

The national discount rate would be guided by the **national climate conversations** and would inform the creation of the **National Climate Partnership Plan**.

The national climate conversations will help people decide what their individual discount rate is. This will assist them in determining the degree to which they value future versus present benefits.

An excellent description of climate discount rates is below:

http://grist.org/article/discount-rates-a-boring-thing-you-should-know-about-with-otters/

Never Normal Coalition

A proposed national coalition whose name is designed to remind us that our **post normal climate** and lives will never be truly normal again over any imaginable period of time. Climate related action must take this unpleasant, even unthinkable, reality into account.

See No Solutions Coalition, Oh Shit Moment

No Solutions Coalition

A proposed organization to combat climate chaos has this somewhat paradoxical name in recognition that climate chaos is an unconventional problem with no real solution. No solution does not mean no action or no hope (or there would be no need to form a coalition).

There will be no end to climate chaos. No WW2 VE (Victory in Europe) day, no equivalent to the famous WW 2 image of the soldier kissing the girl on Broadway, no demobilization from the climate wars.

The mission of the No Solutions Coalition would be to support maximum resilience and readaptation.

See Never Normal Coalition

Oil Repletion Allowance

A proposal for a tax allowance for leaving oil in the ground, providing an incentive for maintaining oil in its unrefined state beneath the surface.

Currently, oil taken out of the ground is given a tax (depletion) allowance, a counterproductive practice since this oil becomes a direct contributor to climate chaos.

See Carbon Retirement, Keep It in the Ground

One Generation Challenge

Many are beginning to see, based on the **carbon budget** and the actual damage that climate change is creating, that the planet has but one generation at most to mobilize a WW2-like effort to transform the world to a **zero-carbon culture.** A One Generation Challenge could become the rallying cry and organizational recruitment tool to enlist millions to personal, civic and national action. Such a campaign could be organized by a broad coalition of civic, corporate and governmental leaders.

The task is vast. Many trillions of dollars of **legacy infrastructure** and fossil fuel facilities will need to be transformed or retired, with **newable** sources of energy constructed in almost unimaginably large quantities. Farming, cities, transport, manufacturing and virtually every other aspect of our way of life will also need to be transformed.

http://www.theclimatemobilization.org/victory_plan

One Hundred Million Climate Scientists

A proposed campaign, to be sponsored by the UN in partnership with a consortium of tech companies, would enlist millions of citizens around the world to gather data to support climate mitigation and adaptation.

One Hundred Thousand Associations for Climate Action

A proposal to mobilize the thousands of trades, fraternal and other associations to support action to reduce climate emissions. As each association's members will be affected by climate chaos in different ways, having many associations harnessing the passions and interests of their members will hasten the country's ability to meet the **One Generation Challenge**.

A look at the websites of a sample of these associations shows very few even mention climate change; thus, the potential is great for association mobilization to make a real difference.

One Percent for Climate

A proposed program that would encourage consumers, corporations and other institutions to direct one percent of all expenditures towards climate adaptation and remediation projects.

Photosyncities

Cities that are designed and managed to maximize the amount of carbon consuming photosynthesis while minimizing CO_2 emissions.

Maximizing parks and green spaces, supporting regenerative agriculture, creating green roofs and walls, substituting renewable energy for fossil fuels, restricting automobile use and encouraging smaller residences are among the many important steps a city can take to become a photosyncity.

See Autotrophic Cities

Physicians for Climate Responsibility (PCR)

A proposal to create a series of professional and other organizations with the goal of supporting immediate climate action. PCR could be one such example. The name is modeled on Physicians for Social Responsibility.

See One Hundred Thousand Associations for Climate Change

Planitect

A proposed new profession.

A Planitect would be trained in all aspects of planetary systems — ecological, biological, economic, architectural and cultural/social. Planitects would work on climate mitigation and adaptation projects, zero emission building design, green community design and related projects.

A Planitect may be considered a kind of planetary physician, perhaps wearing a white coat, but without the stethoscope (except when performing a **carbonoscopy or cognoscopy**).

See Planetarian

Psychopocene

The era of 'new mind'.

A proposal to label the present era the Psychopocene for the influence that our thoughts, feelings and beliefs have on our actions that in turn so profoundly influence the physical world. The **Anthropocene** is the name currently being considered by scientists as the emblem for our era.

Robin Carbon Hood Tax

A proposed small (though some will never acknowledge that any tax could be small) tax on each financial transaction whose proceeds can be used to finance climate chaos **mitigation** and **readaptation** activities.

A tax of this size could raise billions of dollars. Currently, some 40 countries have some sort of financial transaction tax.

The robin carbon hood tax name comes from the redistributionist effects of such a tax, which would fall largely on wealthy individuals and corporations. The original proposal for a financial transactions tax to discourage speculation was developed by Yale Professor James Tobin in 1972.

Simply Zero

A proposed slogan and tagline for a **zero-carbon** movement.

See Zero Now

Stranded Studies

A proposal for a new academic field that analyzes how, why, when and where homes, cities and even nations as well as capital assets, beliefs, practices and behaviors will be stranded or abandoned as a result of climate chaos.

Analyzing what new behaviors, beliefs and actions will emerge after stranding will also be part of this field of study.

What has received most attention in climate circles is the issue of stranded assets. This refers to those assets, particularly fossil fuel related, that may prove to be unrecoverable because of public policy or market conditions. Given the trillions of dollars of value presently attached to fossil fuel reserves, stranding them (or at the least stranding the 80% of them that cannot be burned if we are to have a chance at a habitable climate) is likely to lead to major economic, political and financial shocks.

The mechanisms and implications of asset stranding were examined at one of the first conferences devoted to this issue, held at Oxford University in 2015. It's worth listening to some of the panel discussions.

See Green Assets, Legacy Infrastructure, Legacy Thoughts, Subprime Carbon

http://www.strandedassets2015.org/presentations--videos.html

Ten Thousand Climate Lawsuits

A proposed strategy to accelerate action on climate change that relies upon massive numbers of lawsuits to be filed in jurisdictions all over the US and the world. While the chances of prevailing may vary depending on jurisdiction the publicity value of each suit would be significant.

An International Climate Law Center should be created to encourage and support these efforts by volunteer lawyers throughout the world.

See Children's Climate Crusade

United Climate Way Campaign

A proposal to create a new umbrella charity that will focus on climate chaos related assistance to individuals and charities.

A new organization would elevate the visibility of climate giving and increase the funds targeted to climate chaos **adaptation.**

See Climate Tithing, One Percent for Climate

Universal Climate Design

The design and operation of almost everything in our society — products, services, homes, offices, neighborhoods, hospitals, and more will need to be rethought in a **post normal climate**.

Location, resistance to climate extremes, user needs, availability of goods and services for maintenance and repair and many other characteristics will need to be reexamined to maximize resilience and functionality. This reconceptualization ideally will lead to the incorporation of universal climate design principles into almost everything we design.

A worthy first step would be the convening of an International Universal Climate Design Summit. This meeting of architects, urban planners, industrial designers, and others would be dedicated to articulating a set of universal climate design principles for international review and adoption.

Waltons to the Rescue

It is estimated that the Walton family is worth in excess of $115 billion. Perhaps the Waltons might think about the value of their assets and the security it provides in a world soon to be decimated by climate chaos. That world would probably make Walmart stock close to worthless, and would endanger their lives and those of their families.

What would it take for the Waltons to decide to donate say $113 billion (keeping $2 billion as spare change) to finance climate chaos public education, mitigation research, adaptation projects, climate universities and much more?

They would be acclaimed by the entire world (except the **climate denial triad**). More tangibly this rescue action would improve the chances that their families and friends would survive, as well as many of their namesake stores.

So why aren't **climate hawks** mounting a campaign to persuade the Waltons (or any other hyper-wealthy family) to literally save themselves and the planet?

And why does this entry even have to be written? Why can't the Waltons and every other fabulously wealthy family see what they must do — and then just do it.

See Dirty 90

World Climate Organization (WCO)

A proposed organization that would be responsible for managing and coordinating the world's efforts to control climate chaos. It would be created by and responsible to the United Nations.

It could subsume the **IPCC**, the **COP**, and the United Nations Environmental Program. Creating **climate universities** throughout the world would also be in its portfolio.

The WCO would work closely with the proposed **Carbon Reserve Bank** and **Carbon Cap Commission** to support the administration of cap and trade, a carbon tax or other regulatory measures to limit carbon combustion.

Zero Culture

The creation of a movement to instill a culture that values zero carbon.

A zero-carbon culture would mean that activities are routinely examined for their carbon emissions, recognizing that everyone's survival depends on the sweeping acceptance of drastic limitations on carbon emissions.

Buildings are retrofitted to generate zero emissions, biking, walking or transit takes the place of most driving, a mostly vegetarian diet becomes standard, consumer purchases are limited, and fossil fuels are **intentionally grounded** on behalf of renewables.

Zero Now

A proposed tagline or organizational name for the movement to achieve a **zero culture** right away.

See Simply Zero

Image Credits

Pg. 3 – Albedo, National Oceanic and Atmospheric Administration

Pg. 12 – Carbon budget, carbonbrief.org

Pg. 14 – Carbon capture and storage by LeJean Hardin and Jamie Paynederivative work: Jarl Arntzen (talk) http:// www.ornl.gov/info/ornlreview/v33_2_00/research.htm, CC BY-SA 3.0, https:// commons.wikimedia.org/w/index.php?curid=7268088

Pg. 39 – Paris, December 2015, Photo by author.

Pg. 52 – Geoengineering by Hughhunt (Own work), [CC BY-SA 3.0 (http://creativecom- mons.org/licenses/by-sa/3.0)], via Wikimedia Commons from Wiki- media Commons

Pg. 61 – Kaya Identify by Enescot (Own work) [CC BY-SA 3.0 (http://creativecommons.org/licenses/ by-sa/3.0) or GFDL (http://www.gnu.org/copyleft/fdl.html)], via Wikimedia Commons

Pg. 60 – Keeling Curve, Scrippsnews (https://commons.wikimedia.org/wiki/File:Keeling_Curve_full_record.png), https://creativecommons.org/licenses/by-sa/4.0/legalcode

Pg. 85 – Sunny Day Flooding by B137 (Own work) [CC BY-SA 4.0 (http://creativecommons.org/licenses/ by-sa/4.0)], via Wikimedia Commons from Wikimedia Commons

Pg. 116 – Svalbard Global Seed Bank, Bjoertvedt, Svalbard seed vault IMG 8894, CC BY-SA 3.0

Istockphotos.com – pages viii, 3, 5, 8, 10, 31, 32, 36, 41, 42, 43, 45, 49, 73, 75, 79, 96, 135, 137, 139, 161, 162, 189, 194

Index

C

D

E

F

G

H

I

T

U

V

W

Z

CPSIA information can be obtained
at www.ICGtesting.com
Printed in the USA
BVOW03s0611190817
492323BV00001B/2/P